Climate Policies - Modern Risk-Based Assessment of Investments in Mitigation, Adaptation, and Recovery from Residual Harm

Edited by Gary Yohe and Joel Smith

Published in London, United Kingdom

Climate Policies - Modern Risk-Based Assessment of Investments in Mitigation, Adaptation, and Recovery from Residual Harm
http://dx.doi.org/10.5772/intechopen.1003379
Edited by Gary Yohe and Joel Smith

Contributors
Fabiana Barbi, Gary Yohe, Hippolite Amadi, Indrani Hazel Govender, Julian Schlubach, Leila da Costa Ferreira, Luciana Lima Domingues de Souza, Maliga Reddy, Mpoko Bokanga, Niklas Weins, Paul Clements, Paul L. Woomer, Pedro Andrade Matos, Rajendran Perumal Pillay

First published in London, United Kingdom, 2025 by IntechOpen
IntechOpen is the global imprint of INTECHOPEN LIMITED, registered in England and Wales, registration number: 11086078, 167-169 Great Portland Street, London, W1W 5PF, United Kingdom

For EU product safety concerns: IN TECH d.o.o., Prolaz Marije Krucifikse Kozulić 3, 51000 Rijeka, Croatia, info@intechopen.com or visit our website at intechopen.com.

British Library Cataloguing-in-Publication Data
A catalogue record for this book is available from the British Library

Climate Policies - Modern Risk-Based Assessment of Investments in Mitigation, Adaptation, and Recovery from Residual Harm
Edited by Gary Yohe and Joel Smith
p. cm.

This title is part of the Sustainable Development Book Series, Volume 29
Topic: Climate Change and Environmental Sustainability
Series Editor: Usha Iyer-Raniga
Topic Editor: Yixin Zhang

Print ISBN 978-0-85014-886-2
Online ISBN 978-0-85014-885-5
eBook (PDF) ISBN 978-0-85014-887-9
ISSN 2753-6580

IntechOpen Book Series

Sustainable Development

Volume 29

Aims and Scope of the Series

Transforming our World: the 2030 Agenda for Sustainable Development endorsed by United Nations and 193 Member States, came into effect on Jan 1, 2016, to guide decision making and actions to the year 2030 and beyond. Central to this Agenda are 17 Goals, 169 associated targets and over 230 indicators that are reviewed annually. The vision envisaged in the implementation of the SDGs is centered on the five Ps: People, Planet, Prosperity, Peace and Partnership. This call for renewed focused efforts ensure we have a safe and healthy planet for current and future generations.

This Series focuses on covering research and applied research involving the five Ps through the following topics:

1. Sustainable Economy and Fair Society that relates to SDG 1 on No Poverty, SDG 2 on Zero Hunger, SDG 8 on Decent Work and Economic Growth, SDG 10 on Reduced Inequalities, SDG 12 on Responsible Consumption and Production, and SDG 17 Partnership for the Goals

2. Health and Wellbeing focusing on SDG 3 on Good Health and Wellbeing and SDG 6 on Clean Water and Sanitation

3. Inclusivity and Social Equality involving SDG 4 on Quality Education, SDG 5 on Gender Equality, and SDG 16 on Peace, Justice and Strong Institutions

4. Climate Change and Environmental Sustainability comprising SDG 13 on Climate Action, SDG 14 on Life Below Water, and SDG 15 on Life on Land

5. Urban Planning and Environmental Management embracing SDG 7 on Affordable Clean Energy, SDG 9 on Industry, Innovation and Infrastructure, and SDG 11 on Sustainable Cities and Communities.

The series also seeks to support the use of cross cutting SDGs, as many of the goals listed above, targets and indicators are all interconnected to impact our lives and the decisions we make on a daily basis, making them impossible to tie to a single topic.

Meet the Series Editor

Usha Iyer-Raniga is a professor in the School of Property and Construction Management at RMIT University. Usha co-leads the One Planet Network's Sustainable Buildings and Construction Programme (SBC), a United Nations 10 Year Framework of Programmes on Sustainable Consumption and Production (UN 10FYP SCP) aligned with Sustainable Development Goal 12. The work also directly impacts SDG 11 on Sustainable Cities and Communities. She completed her undergraduate degree as an architect before obtaining her Masters degree from Canada and her Doctorate in Australia. Usha has been a keynote speaker as well as an invited speaker at national and international conferences, seminars and workshops. Her teaching experience includes teaching in Asian countries. She has advised Austrade, APEC, national, state and local governments. She serves as a reviewer and a member of the scientific committee for national and international refereed journals and refereed conferences. She is on the editorial board for refereed journals and has worked on Special Issues. Usha has served and continues to serve on the Boards of several not-for-profit organisations and she has also served as panel judge for a number of awards including the Premiers Sustainability Award in Victoria and the International Green Gown Awards. Usha has published over 100 publications, including research and consulting reports. Her publications cover a wide range of scientific and technical research publications that include edited books, book chapters, refereed journals, refereed conference papers and reports for local, state and federal government clients. She has also produced podcasts for various organisations and participated in media interviews. She has received state, national and international funding worth over USD $25 million. Usha has been awarded the Quarterly Franklin Membership by London Journals Press (UK). Her biography has been included in the Marquis Who's Who in the World® 2018, 2016 (33rd Edition), along with approximately 55,000 of the most accomplished men and women from around the world, including luminaries as U.N. Secretary-General Ban Ki-moon. In 2017, Usha was awarded the Marquis Who's Who Lifetime Achiever Award.

Meet the Volume Editors

Gary Yohe holds a BS degree in mathematics from the University of Pennsylvania and a Ph.D. in economics from Yale University. Since turning his attention to climate change in 1981, he has contributed 10 books, 175+ research articles, and 75+ opinions to our understanding of climate risk and humanity's response options. He served as a senior member of the United Nations Intergovernmental Panel on Climate Change from 1995 through 2018 and shared the 2007 Nobel Peace Prize. He was the Vice Chair of the 2014 Third National Climate Assessment for President Barack Obama and a charter member of the New York City Panel on Climate Change for Mayor Michael Bloomberg. He continues to serve as Co-editor-in-Chief of Climatic Change.

Joel B. Smith is an expert with over 35 years of experience researching climate change vulnerability and adaptation. He led the first national assessment of climate change impacts, the U.S. Environmental Protection Agency study of climate change impacts in the late 1980s, and the first global assessment of climate change impacts, published by Cambridge University Press in 1995. He served twice as a Coordinating Lead Author for the Intergovernmental Panel on Climate Change, where he conceived of and led the development of the synthesis of climate change impacts (a. k. a., "Burning Embers"), which has been updated in subsequent Intergovernmental Panel on Climate Change (IPCC) reports. He has published over 70 journal articles, edited books and book chapters, and written two dozen reports on climate change vulnerability for government and research organizations. Mr. Smith worked for the Environmental Protection Agency (EPA) and has been a consultant for most of his career. He taught at universities on climate change and served on several advisory boards for the U. S. government and the United Nations.

Contents

Preface

When John Holdron spoke for the President during his 8-year tenure as Science Advisor to President Obama and Director of the White House Office of Science Technology Policy, he was fond of emphasizing some variant of a concise summary of humanity's options with regard to changes in climate-related risks born of dangerous human interference with the climate system itself. "When it comes to responding to climate policies", he would say and I concur that, "we have three choices – abate, adapt, or suffer".

Abatement is called mitigation by those involved in climate science and climate action – actions designed to reduce emissions of greenhouse gases. Adaptation involves making investments that can ameliorate some of the impacts that persist under all feasible mitigation pathways made with the understanding that no adaptation will be effective 100% of the time. Suffering will, therefore, always occur as a residual to any combination of these two active policy actions; that is to say, humanity cannot escape all of the impacts of rising temperatures, especially the most extreme possibilities that are known unknowns. Suffering is the "accept a tolerable level of risk" option that informs the level of investment in the other two choices. To be precise, setting the level of what is and is not tolerable is really the third policy option.

Earlier in 2007, the Intergovernmental Panel on Climate Change had achieved consensus support for a finding that fundamentally changed the way that the science community and their clients in the decision-making realm viewed the climate change challenge: "Responding to climate change involves an iterative risk management process that includes both *adaptation and mitigation* and takes into account climate change damages, co-benefits, sustainability, equity, and attitudes to risk" (*my emphasis*; Summary for Policy Makers of the Synthesis Report for the Fourth Assessment Report in 2007) [1].

These worlds provide us access to a productive way of organizing our thoughts and assessments with regard to the relative values (positive or negative) of various mixes of the three options because they make the concept of risk the connecting tissue. At its very core, risk is simply the product of likelihood and consequence. Mitigation holds the potential to reduce the likelihood of damaging events (or at least delay them). Well-informed adaptation can reduce the potentially dire consequences of the warming that remains after abatement has done what it can. And perhaps investments in resilience and recovery programs can ameliorate some of the residual suffering.

This collection of chapters is organized in sections that reflect Holdren's three broad categories of climate policy options. All apply a risk-based perspective to discussions and assessments of what we do and do not know and, thus, what we can or perhaps cannot really do. Each chapter includes, in their concluding remarks at least, some attempt to identify critical gaps in our current understandings of the coupling of the climate and human systems on earth and, by continuation, elaborate on constraints on our confidence in relative efficacies that we project in our policy deliberations. Specifically, efficacy judgements take account of at least one of the IPCC metrics: net climate change damages, co-benefits and costs of policies, measures of sustainability (of systems and policies), equity calibrated in various metrics of human welfare security, and the degree to which people, communities, sub-national governance bodies, nations, and international institutions are averse to risk.

Gary W. Yohe
Huffington Foundation Professor of Economics and Environmental Studies,
Wesleyan University,
Middletown, CT, USA

Joel B. Smith
Independent Researcher,
USA

References

[1] Yohe G. How to think about climate change responses: on organizing one's thoughts. In: Lackner M et al. editors. Handbook of Climate Change Mitigation and Adaptation (Third and Fourth editions). Springer Nature, Springer Science+Business Media, LLC; 2021 and 2024. Availble from: https://link.springer.com/content/pdf/10.1007%2F978-1-4614-6431-0_102-1.pdf and https://doi.org/10.1007/978-1-4614-6431-0_102-1

Section 1

Mitigation and the Reduction of the Likelihood of Dangerous Climate Impacts

Chapter 1

Climate Change and LMIC Neonatal Care at a Crossroad: Self-Driven Research for Policy-Actionable Adaptation and Mitigation

Hippolite Amadi

Abstract

The challenge of global warming is real with its consequent risks which exacerbate morbidity and mortality of LMIC neonates, especially in tropical hot climates such as West Africa. High grid and fossil-fuel energy consumption in industries and healthcare contribute to climate-change. This includes neonatal healthcare where continuous power consumption is needed for the operation of equipment for interventions that keep neonates alive. It is morally unjustifiable to stop the fossil-fuel generator when the life of the newborn is dependent on its operation. The climate dilemma is that the more we consume carbon energy, the worse global warming gets, and the worse the fatal neonatal hyperthermia due to increased global warming. Here lies the crossroad that this Book Chapter seeks to explore. Therefore, efficient climate policies may need to focus more on carefully synthesized bidirectional solutions that must both save the neonate from the harsh climate and also save the climate from neonatal care activities. Neonatal care is just one of numerous aspects of industrial healthcare that could be viewed in a similar manner.

Keywords: neonatal devices, global warming, evening-fever syndrome, climate policy, LMIC

1. Introduction

Emerging climate data of 2024 has revealed further evidence of the continuing global warming trend in 15 of the last 16 months prior to November 2024 [1]. The Copernicus Climate Change Service (C3S) declared that based on the data of the first 10 months of 2024, it was certain that 2024 would be the warmest year on record and the first year of more than 1.5°C above pre-industrial levels [1].

Climate-change narratives might seem a sophisticated discourse on the global stage with the current high-powered Climate Summits across the continents of the world, spear-headed by the United Nations and various Committees around the world; however, for an ordinary uneducated man in many low- and middle-income countries (LMIC) such as Nigeria, faraway in their villages, there are evident realities of harsh climate upon everyone. Ordinary people across the villages of Nigeria can only be thankful if the 'spirits of their ancestors' could arise to ameliorate the debilitating harsh impact of changing climate upon them—from unbearable environmental heat to frequent flooding experiences across the nation. The climate and environmental heat conversation have, therefore, become extant issues of discourse in every local language and across every age group in many LMICs, and deserving of urgent intervention. The reality of our current situation may seem like a sudden change to some people; however, the signs of the impending dangers have been with us for a long time and may have remained sufficiently unchallenged.

Recent reports in social media and conventional News Media across Nigeria, for example, have seen an increase in the outcry of youths and adults—complaining of difficulties in coping with the high environmental heat. The extended dilemma for any scientific mind would wonder how much more difficult this would be for little children and newborn babies, who are unable to interact on social media platforms. The health of humans has been at the forefront of the climate conversations of the United Nations—hence, LMIC villagers are not immune to this. Furthermore, as an 'electric power-guzzling sector', healthcare is a major contributor to global high carbon footprint, which has been said to be driving the present global warming. Therefore, human adaptation and resilience, and mitigation of the current climate issues become a necessity and must not be neglected by policymakers. This includes the unwellness and indirect contributions of LMIC newborn babies who are unable to participate in the discussion of these issues with adults. This also involves the need to seek the best approach of communicating the local LMIC knowledge of healthcare impact in the global climate discourse as has been recently amplified [2].

1.1 Objectives

The objective of this article is to highlight my previous and current scientific investigations, interventions, and the development of technological mediations for neonatal climate-change resilience and adaptation, and the innovation of policy-adaptable technologies for the mitigation of the contributions of neonatal medicine to climate-change in Nigeria, and by extension, the LMICs.

2. Neonatal safety and adaptation techniques for climate resilience

2.1 Overview

The need to develop ideas to innovate technologies that could assist humans, and especially newborn babies, to adapt to rising global temperatures might not materialize unless the underlying pathways are discovered and subdued. Africans and the other underdeveloped climatic regions of the world have been suffering from rising environmental temperatures for a long time but without many investigations of how this had affected newborn morbidity and mortality of African neonates. The literature has rather been dominated by neonatal hypothermia (cold stress) but has little about neonatal hyperthermia (heat stress), which is relatively more prevalent in Africa than the colder regions of the world [3]. Hence, little was known about all causes of neonatal hyperthermia, including how climate effects might have been partly responsible for some hyperthermia-associated morbidities and mortalities.

The high fever accompanying neonatal hyperthermia in Nigeria was always blamed on 'possible disease process', and hence, often ending up on prescription of prophylactic antibiotics whilst laboratory results were being awaited. The climate was never suspected, investigated, or implicated for neonates presenting with hyperthermia in Nigeria until my studies of 2010 was published in the literature [4]. The lack of adequate knowledge of climate-change on the wellbeing of the Nigerian neonate, has thus plunged many to the devastations of wrong antibiotic prescriptions and administration, and perhaps, eventual death.

2.2 Discovery of neonatal evening-fever syndrome (EFS): Ilorin 2010

I served as a Visiting Consultant Researcher at the neonatal intensive care unit (NICU) of the university of Ilorin Teaching Hospital in Kwara State Nigeria between 2005 and 2011. During one of my visits in 2009, a compelling phenomenon drew my attention. This occurred during the extremely hot months of 2009.

The splash-back of one of the wash sinks in NICU was finished with a set of beautifully decorated ceramic tiles. I had loudly admired the beauty of these tiles the previous day as I worked with one of the nurses. However, whilst I was consulting the next day, I discovered that one of the tiles had cracked in full length. I was displeased and requested an answer on who it was that broke the tile. The ward-manager replied that 'self-cracking of tiles in the NICU' was an acknowledged occurrence and happened from time to time, especially during hot weather; hence, no one was responsible. I left the Unit very much troubled as to what was going on. If the weather had done this to the tiles in the NICU, what else is playing out and how could this be affecting the incubators and the neonates? I questioned myself in my thoughts. Hence, I ordered and designed a study to scientifically investigate how the incubators were responding to whatever it was that cracked the tile. This research lasted one full year revealing the compromise of incubator functions owing to high sunlight intensity [4].

Therefore, I hypothesized that the neonates who were being nursed in the incubators and cots in Nigerian NICUs could be suffering from unknown or undisclosed capacitance-effects of high sunlight intensity in the nurseries. There was an urgent need to investigate this with a specially designed study protocol. An ethical assessment application was submitted, and approval was obtained for this study at the Federal Medical Centre Nguru—situated at extremely hot climate region of north-eastern Nigeria (Yobe State) in 2011. The preliminary results revealed a disturbing high neonatal morbidity owing to climate-induced neonatal hyperthermia, which incidentally and characteristically disappeared after sunset each day [5]. I codenamed this phenomenon "The Evening-fever Syndrome (EFS)" as submitted to the World Health Organization global forum on medical devices in 2013 [6]. The need for possible solutions became inevitable.

2.3 The politeshield antidote – A pattern: Nguru 2013

This is a unique pattern of climate-mitigating Nursery Building Features as the antidote for ameliorating neonatal condition of Evening-fever Syndrome (EFS) in the Nigerian neonate as was discovered in my earlier studies in 2012. Nursery building in hot climates such as Nigeria may be contributing to the process of unintended overheating of newborn babies—iatrogenic neonatal hyperthermia. This is a condition of periodic fever that overcomes the neonate in the mid afternoon and early evening hours, which has been studied to be climate-induced and propagated via heat transfer mechanisms through the walls of the nursery building. This is otherwise called neonatal evening-fever syndrome as I code-named it in 2012.

I hypothesized that the process of transmitting the climatic heat of high sunlight intensity to overheat a neonate inside a nursery building may only be truncated via specialized building features that are designed to mitigate the EFS.

3. Politeshield design background

One of the major causes of high neonatal mortality rate (NMR) in Nigeria is extremely poor neonatal thermoneutral support. Many clinicians and nurses understand the concept of neonatal hypothermia owing to dropping room temperatures of a nursery hall, and practitioners' subsequent search for whatever available equipment or technique to keep such newborn baby warm. The other thermoneutral extremities—hyperthermia and rising room temperatures of the nursery hall—are not well-understood. Many clinicians suspect a disease process during neonatal hyperthermia. However, neonatal hyperthermia can also result owing to a climatic heat capacitance effect whereby the external walls of the nursery building absorb intense heat from sunlight during the day, conducts this through the wall, and radiates it into the room during the evening, hence, uncontrollably overheating the neonates. The excessive overheating that results from this is not a disease process and can only be prevented with a nursery building design that can feature concise mitigants that target this high climate effect. Nigeria is geographically located within the equatorial tropical belt, having a high climate activities of high sunlight intensity all year round. Such impact of the regional climate, especially in the absence of air-conditioning units, exposes the nursery buildings to become thermally charged by the sun, and subsequently unleash the devastating hyperthermia on the neonates leading to high morbidity, or even deaths. Most nursery facilities in Nigeria and other LMICs cannot afford air-conditioners; albeit this may be unacceptable in the present drive against greenhouse gases [7]. The common nursing practice for ameliorating this condition of sudden neonatal hyperthermia in the northern part of Nigeria often involves soaking a face towel in cold water to sponge the baby from head to foot. This practice can lead to a devastating lethal condition called "neonatal thermal shock" [3].

Hence, the need existed to propose, design, and develop a specific nursery building pattern with notable mitigating features against climate-induced neonatal hyperthermia, which could become a policy gold standard for LMIC neonatal facilities for adaptation, most especially in the northern regions of West Africa, where ambient temperatures often reach 47°C during the hot seasons.

3.1 Concept

My concept of the features of the EFS-antidote design—which has since been in use in Nigeria for over 10 years—stems from the studied mechanisms of the propagation of the EFS as explained below.

DOI: http://dx.doi.org/10.5772/intechopen.1008563

i. It is natural for the sun to rise with quick attainment of high sunlight intensity in Nigeria.

ii. It is also natural for the striking sunlight to heat up the exposed walls of the nursery building depositing much of its heat energy.

iii. It is natural for the walls to conduct the heat through to the inner surface of the nursery walls and hence release the heat through radiation to the babies inside the nursery, thereby initiating the EFS against the neonates.

Therefore, my concept was to create a design with features that could truncate this chain of events, thereby limiting the delivery of the sunrays' deposited heat to the neonate being nursed in the open nursery hall.

The concept seeks to achieve a naturally cooled nursery room without the need of an air-conditioner unit through three mechanisms: by

i. limiting the radiant heat transfer from the charged outside-aspect of the nursery wall into the inner aspects of the nursery building as depicted in **Figure 1** (left)

ii. naturally tapping any environmental coolness into the inner aspects of the nursery building (such as lowering the inner floor below the ground level) for

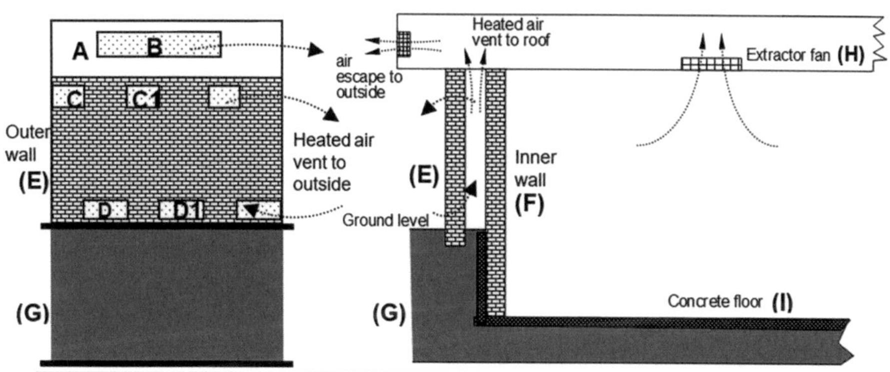

LEGEND
A – the nursery building parapet
B – heated air vent to the intra-roof air space, covered with rodent-mesh sheet
C, C1 – outer wall top aspects of heated air vent (outlet) from the inter-wall air space, wall opening covered in rodent- mesh sheet
D, D1 – outer wall bottom aspects of air vent (inlet) into the inter-wall air space, wall opening covered in rodent-mesh sheet
E – outer aspect of the double wall: On the right diagram—cross-sectional view showing interwall air space; and on the left diagram—elevation, showing air vents
F – inner aspect of the double wall showing its cross-sectional view
G – underground soil
H – ceiling mounted extractor fan aiding heated air escape via the roof space
I – concrete flooring of the base and sides of the pit

Figure 1.
Air-lagged double wall technique as modified from Amadi et al. [8]. (Left – a view of the outer wall of the building. Right – a cross-sectional view through the double wall to reveal the air gap for natural convention, innovated to truncate the transmission of heat across the inner wall).

producing a thermal counter effect against any heated air within the nursery as depicted in **Figure 1** (right)

iii. integrating a heat-exchanger module across the inner walls of the nursery: This is a single or double channels of crisscrossing 10 mm copper pipe around the perimeter of the nursery. The heat-exchanger pipe runs continuously from its entry port, where it is connected to an overhead water tank supply, snaking top-to-ground level continuously across the perimeter of the inner walls until its exit port, from where it discharges its relatively warmer water into an underground tank (**Figure 2**). The natural water pressure-head enables cool water in the overhead tank to travel via the pipe, absorbing the heat already transferred to the copper pipe from the walls through conduction across the aluminum-sheet overlay (component D, **Figure 2**). Hence, the discharged water at the exit of the heat-exchanger pipe would be relatively warmer than the temperature of the water at the inlet port. This aspect of the design targets to remove the heat already entrapped on the inner walls of the nursery building before this radiates into the open space of the nursery impacting the neonates.

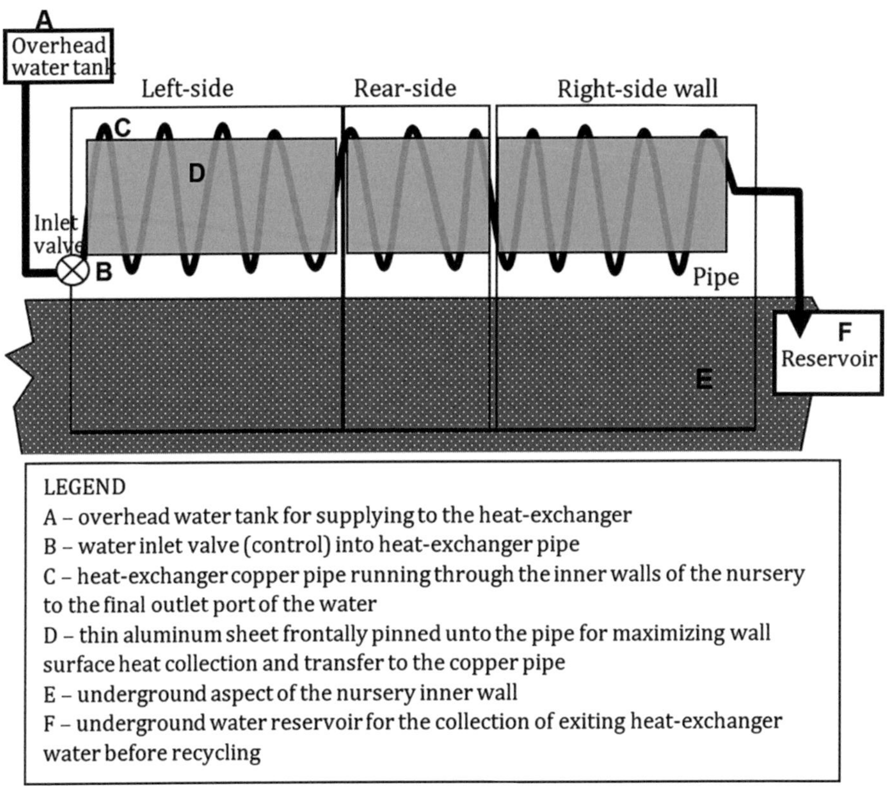

Figure 2.
Schematic representation of the unwrapped inner walls showing the Heat-exchanger copper piping through the walls in the trial laboratory at FMC Nguru, Yobe State - from Amadi et al. [8]. (Overhead tank delivered water through an inlet valve. Circulation was ensured automatically through the positive pressure-head under gravity created by the relative height of the overhead tank. Cool water in the underground reservoir was pumped back to the overhead tank when required. This process removes the heat already trapped in the inner walls before it radiates into the open nursery space).

3.2 Brief description of drawings

An effective tropical climate-resilient nursery should be able to stop, divert, or minimize the heat transfer process from the outside-building environment to the inside environment of the nursery building as demonstrated in my design, shown in **Figure 1**.

The building system was so-created, hence, diverts and truncates the heat transmission across the walls, thereby limiting the reach of the negative impact of high climatic temperatures on the nursery inmates [8].

3.3 Summary

This is the first of this kind of building features for the protection of neonates from the effects of climate-change and evening-fever syndrome (EFS). The innovation is adoptable by LMIC policymakers and is suitably built to operate in the urban and rural parts of equatorial regions of the world where ambient temperatures soar too high necessitating the high demand for air-conditioning units which further adds to the complications of global warming. This is a proven technique of natural cooling of a Nursing-Bay which has been shown to naturally keep the Hippolite-FMC Nguru laboratory Nursing-Bay at maximum of 32°C during an outdoor environmental temperature of up to 46°C (**Figure 3**) [8].

3.4 Nigeria adoption, versions, and policy impact

In the last 10 years, the clinical knowledge and diagnosis of evening-fever syndrome (EFS) have received the attention of the Nigerian Society of Neonatal Medicine and widely considered during practice across the Special Care Baby Units (SCBUs) nationwide in Nigeria. This has enabled clinicians to query and rule out the climatic-effect in any presenting neonatal fever before the prescription of antibiotics or arrival of lab results—which often arrive late. This awareness is revolutionizing neonatology-based clinical decisions for patient care. This has also widened the knowledge of neonatologists and their suggestive inputs during the planning and design stages of new nursery buildings in Nigerian hospitals.

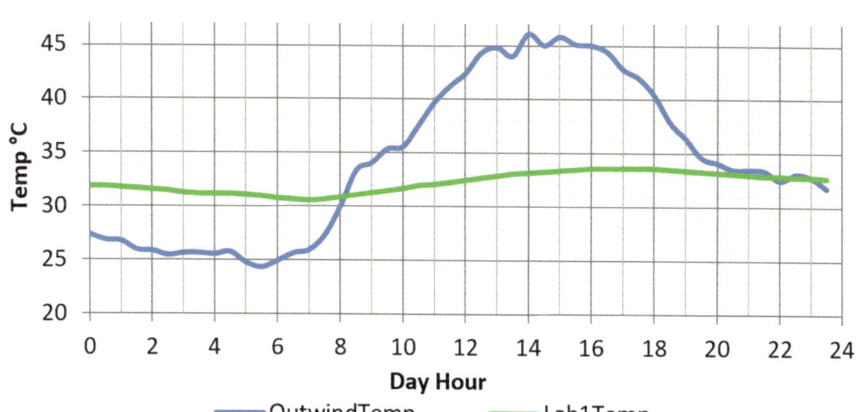

Figure 3.
Typically captured performance of Lab-1 on a hot day, 7th April 2013 with a high of 46·1°C. ('OutwindTemp' is the temperature of outdoor environment; 'Lab1Temp' is the ambient temperature of the Nursing-Bay).

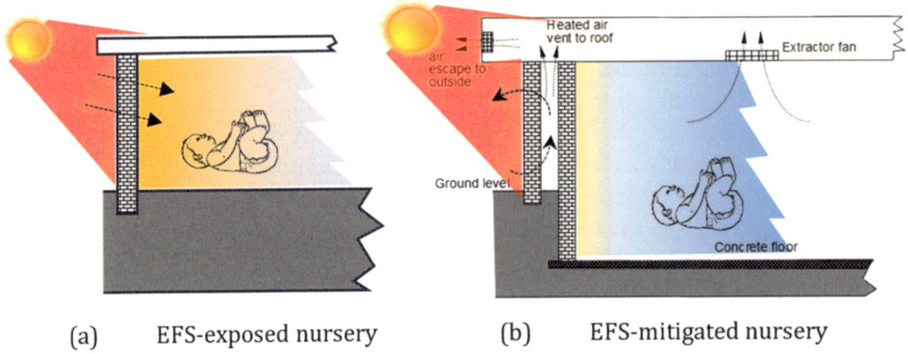

Figure 4.
EFS-exposed and EFS-mitigated nurseries (Diagrammatic expression of climatic heat impact on a neonate in a typical nursery – EFS exposed, as compared to a climatic EFS mitigated nursery).

The original trial of this innovation between 2011 and 2013 presented two laboratories—LAB1 with all features for a freshly built nursery and LAB2 with fewer upgrading features for an already existing nursery. These versions have been discretely adopted and implemented at various places across Nigeria, some preferring the adoption of Lab-2 over Lab-1 for minimization of policy risks on cost.

In a representative large-scale policy application, the Yobe State government of Nigeria initiated a unique Statewide healthcare delivery project relying on the lessons learnt from the EFS neonatal laboratories. Hence, delivering their own version of the Neonatal Rescue Scheme [9], and constructing new regional complexes for newborn care after the pattern of the EFS Lab-1 features as showcased in their 2023 documentary [10]. The relevant aspect of this documentary can be accessed from this YouTube clip – https://youtu.be/5Vvc6Cvc4So. Some other new building complexes adopted this as showcased in "Inside Africa" documentary of CNN International [11].

3.5 Conclusion

The deadly effects of climate-induced EFS could be minimized in tropical LMICs if the choice of the site and positioning of the nursery apartment within the neonatal building complex is made with full evaluation of the predictable interactions and impacts of sunrays as the sun rises from the eastern cardinal direction and sets at the western. The neglect or ignorance of EFS mitigation would lead to daily neonatal hyperthermia, complications from needlessly administered antibiotics, and neonatal thermal-shock from practices that attempt to rapidly force-down the neonate's body temperature, which could lead to death [3, 12]. The EFS discovery and my innovation of its antidote have been a bold response in pushing back one troublesome aspect of climate-induced morbidity on humanity (**Figure 4(a)** and **(b)**, respectively).

4. Towards zero-carbon: Could healthcare climate-damage be mitigated?

4.1 Overview

It was a sudden realization—as I was concluding the 2010 Ilorin study on the impact of high sunlight intensity [4]—that the climate could be inflicting enormous amount of damage on the Nigerian neonate. This ignited in me a long-drawn storm of reflections on how the age-long activities of humans could have adversely compromised the conducive climate our Nigerian ancestors enjoyed. This consciousness became a driving force for my subsequent creations and implementation of climate-friendly neonatal technologies in my Nigerian neonatal practice.

The rising global temperatures—if unchecked—with its possible impacts on our reproductive physiology leave us with the uncertainties of what could become of humanity's ability to reproduce quality offspring in the future. As I built fences to protect the Nigerian neonate from devastating climate-change consequences such as the EFS, I was compelled to join in the footsteps of other scholars studying to understand why our climate was changing. I realized that our high carbon footprint as humans, which has negatively impacted the climate, has not exonerated our other activities to keep the sick neonate alive. The enormous use of electric power in pharmaceuticals and entire healthcare delivery systems—including neonatology which I practiced—might have been a contributor to the damage to our climate. Therefore, I concluded that I must find ways of cutting back on the carbon contributions of the neonates in my African practice. Effective actions could be radical, extraordinary, and particularly unconventional—hence, I needed enormous determination and courage to significantly push any boundaries.

The major interventions in neonatal care are implemented with machines that require to operate continuously using mains electricity—national grid power or fossil fuel generators—for as many days as the neonate remains on admission. This constitutes a huge amount of environmental carbonization. Therefore, I felt the urgent need to begin to think of how I could strike a balance between keeping our neonates alive and saving the planet for them as they survive to grow up to meet the devastation we have caused. This is a serious matter as neonatology in the present context could, therefore, be classified as an electric power-guzzler in healthcare terms owing to the numerous electrically operated life-support machineries and gadgets that MUST run simultaneously and continuously for several weeks – nonstop – just to keep the neonate alive. These are the incubator, the ventilator, the phototherapy machine, the patient monitor, etc.—electric power guzzling machines which must necessarily run without a break for 2 to 4 weeks or more until the neonate is 'out of the woods' or goes to 'the world beyond'. We need to keep them alive, but we even more need to preserve the planet for them.

In the earlier years of my African practice, I was embattled with the well-known Nigerian epileptic power issues. The unwell neonate cannot do without power, such that in the sudden failure of grid energy as is often the case in Nigeria, the caregiver must necessarily force some hundreds of kilograms of carbon into the atmosphere, continuously running fossil fuel generators—where this is available—to keep the neonate alive. The balance must lie in lowering the carbon-footprints from the use of all these essential technologies for neonatal care by innovations that would mitigate how the interventions contribute to climate-change. Hence, I faced another challenge of innovating methods of tweaking the operational systems of existing neonatal devices, developing tailored designs, and creating alternative green energy techniques to make the neonatal devices and facility-lighting to become less dependent on mains electricity. The tweaked technologies must be efficient in power consumption management to ensure 24 hours/day reliability and sustainability. The journey to achieving zero-carbon emission in my neonatal practice seemed impossible, but this had to start with a review of my existing technologies.

4.2 Decarbonization of the power-banking-system

Leading up to the year 2009, I was simultaneously engaged with over 12 tertiary and university teaching hospitals across all regions of Nigeria as a Visiting Consultant, supporting their neonatal thermoneutral research and interventions. Sudden power failures in Nigeria were rampant. The often-long waiting time in-between fossil-fuel power change-over at the Special Care Baby Units (SCBU)—where I practiced—retarded the progress of thermoneutral care leading to the loss of lives. Hence, I introduced the power-banking-system (PBS), which used 48-volts 800AH gang of batteries and 4 KVA load-capacity inverter modules to bank-up mains electric power to later utilize this to run the incubators and facility lights during the power-failure waiting time. The PBS functioned like a UPS that could deliver up to 4 hours of power support—a great deal of welcomed development at the time, especially by 2014 when I had perfected this art, applying it at many SCBUs across the country—in southern and northern Nigeria. This became a springboard for my new challenge requiring a major shift in the source of power for all my neonatal systems at the Nigerian SCBUs. My initial small climate action was to end the recharging of the PBS with mains electricity. It was almost seamless to shift from mains power supply to a gang of photovoltaic cells (solar panels) as the means of re-charging the PBS batteries—a successful move in reducing the neonatal carbon footprint by that small amount.

The conventional technique of inverter power installation in Nigeria involves connection of the panel-battery-inverter assembly as a standby power supplier in the event of grid power failure. This technique has limitations that often create practice frustrations as was observed across my network of SCBUs applying this. Some centers explained their frustrations for the lack of sustainability of this power option, including the high cost of frequent deep-cycle battery replacement. Other centers complained of the short lifespan of the standby inverter power supply after a full charge during sunshine, which could only support facility lighting and powering of the essential neonatal machines at night for less than 4 hours before the SCBU returned to utter blackout for the rest of the night. There were explainable technical reasons for this short lifespan, including inefficiency of the installation technique that was often applied. I embarked on a scientific investigation for a more efficient way of making the banked energy stretch further to cover the power needed for an entire night shift. Hence, in my resulting publication, I proposed and tested the separation of the duo of 'facility lighting' and 'sockets powering' into two independent systems when using renewable energy as the main source of powering the SCBU [13]. The proposal demonstrated how the Neonatal Unit could be powered by two independent solar panel gangs, one supplying 12-volts to its charge controllers for facility lighting purposes and the other supplying inverter-able 48-volts for powering the equipment, respectively. The new solar-based PBS technology has been successfully applied at seven notable Nigerian hospitals, including the SCBU-1 of Federal Teaching Hospital Owerri, the Amina-Centre of General Hospital Minna (northern Nigeria) and the Neoroom of the Calabar Women and Children Hospital (southern Nigeria), enabling these hospitals to often operate on 'all night' solar-banked PBS-power until daylight returns the next day when the system returns to the regular supply of harvested solar power in real-time to operate the neonatal equipment.

4.3 Invention of the PLB: A solar-based technology to decarbonize facility lighting

It is literally impossible to do modern healthcare delivery without power for lighting the environments of the healthcare facilities especially during the night. It is required, by standards, that a healthcare facility does not practice in the dark—adequate high illumination of environment lighting is a 'non-negotiable' in neonatal care delivery. The enormous air pollution from fossil fuel generators to keep the nursery unit's high wattage bulbs alight endangers health as well as contribute huge amount to climate-change, hence requiring some scientific mitigation.

Independent lighting system: In Ref. [12], authors reported that typical inverter power in Nigeria is wired into the existing nursery power distribution lines using 'switchover gear' controls via chosen fuse-lines. This technique enables them to select specific fuse-lines for power sockets and gangs of lighting points to be energized by the inverter. However, all the conventional high voltage appliances along the prioritized lines are automatically enabled. "Therefore, the inverter system is subjected to supply energy to all installed conventional light bulbs on the prioritized fuse lines", most of which are rated up to 100 watts [12]. I had previously reported in the literature—from my experience—that the inverter system installed in this manner would readily get drained of its battery reserve in quick time, and that low wattage bulbs, such as 5 to 12 watts had been tried but these yielded insufficient illumination intensity for critical aspects of clinical works, such as 'neonatal line-setting' in the night. Therefore, we innovated the 'polite-light-bank (PLB)' system, which was a successful response to the need of making the solar power option to sufficiently last an entire night with high intensity of illumination throughout [13]. The PLB was a separate low-voltage wiring—independently distributed across the walkways and rooms of the nursery building, applying Light-emitting-diode (LED) technology. Hence, the lighting system so-demonstrated was a standalone 12-volts system, which relieved the inverter system from the power-draining high-wattage light bulbs. The application of the PLB technology as an essential system for high illumination lighting of the neonatal facility, hence, presented a major shift and a win for my decarbonization agenda in neonatal healthcare delivery (**Figure 5**). The applying facilities gave an average overall satisfaction rating of 98% in their performance assessment over 4 years of applying this lighting technology [13].

4.4 Solar-adaptation, decarbonizing existing neonatal devices and new inventions

The PLB technology was one huge neonatal decarbonization success that I needed to exploit further for the Nigerian and LMIC neonates. Its low-voltage and low-power consumption characteristics—requiring no intermediary inverter module—make it an extremely low-cost item of essential technology for neonatal care in addition to its zero-carbon pollution quality. Therefore, beyond 2020, I began to research and trial how the PLB technique could simultaneously power some essential neonatal devices that otherwise rely on grid or fossil-fuel power. The conscious elimination of the 'inverter module' where possible, has the general advantage of lowering maintenance and operational costs of the power system—inverters cost much to acquire and are prone to unavoidable and avoidable failures that could shutdown life-support devices until maintenance engineers and required spare parts arrive and repairs carried out. These could take considerable amount of time—days to weeks—to happen with the consequences of loss of lives.

Individualized power-buffer station: The "individualized power buffer station" was conceptualized to create independent low power reserves for all applicable neonatal devices in the SCBU. Typically, in our concept, a pre-processed energy from the PLB solar panels is split and channeled to a few "buffer stations", strategically located

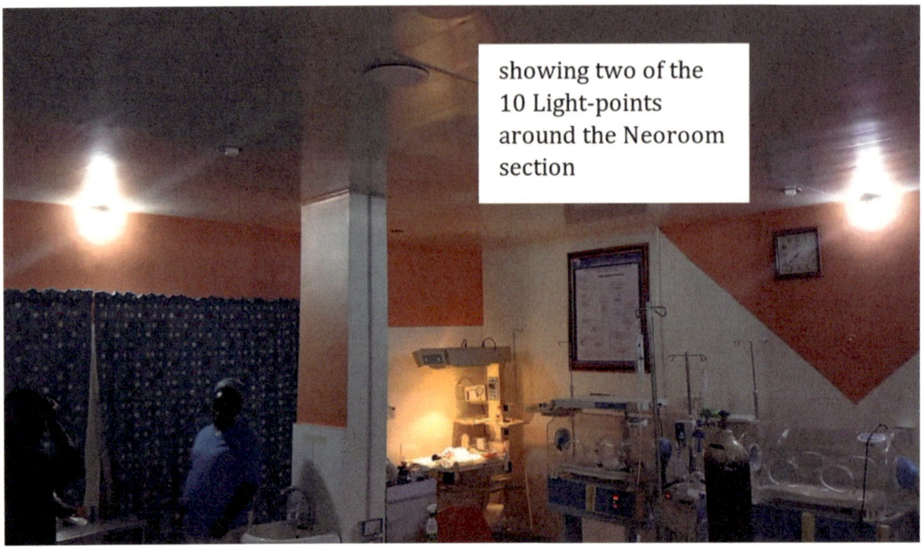

Figure 5.
Neoroom at Calabar Women & Children Hospital, Calabar at night-time. (Situated in the outskirts of Calabar, the neoroom operates on zero-carbon lighting system and 100% reliant on low-voltage banked solar energy all night long).

around the inner nursery ward where two incubators could access the same buffer station for energy without being moved around. Unlike the PBS, which is fully installed in a separate 'power room' in the nursery complex, an individual buffer-station is attached to the equipment or located near the devices within the room [12]. The buffer stations are 12-volt 150AH-battery systems linked to the 12-volts solar-panel gang supply of the facility's PLB and recharged via a dedicated MPPT charge-controller, each delivering outputs of 12-volts DC via cigarette lighter sockets, and 230 volts AC via a dedicated 2.0-3.0 KVA power inverter.

Relying on the successful creation of power outputs from the buffer-station, applicable neonatal devices were re-manufactured or re-engineered by creating a power interface that enabled them to operate on the preserved solar-energy. Some of the decarbonized neonatal devices are described below:

1. The innovation of decarbonized incubator & resuscitaire (**Figure 6a**)—being the creation of the green-energy versions for my >20 years old popular invention – the RIT technology [14, 15]. In the late 1990s, Nigerian tertiary hospitals were notorious of littered carcasses of obsolete incubators—abandoned in rooms, workshops, and along walkways, breeding rodents and mosquitoes and causing environmental pollution. Beginning from 1996—when I took notice of this trend—I began a long-stretched study on possible innovations that could minimize this environmental pollution trend and simultaneously reuse the same rejects to enrich neonatal care in Nigeria. The concerted work was rewarded with my invention of the recycled incubator technology (RIT) in 2003 and the use of this technique across the entire country to reclaim hundreds of dysfunctional incubators back into functional status—up to the end of MDG4 in 2015. My climate concerns beyond 2015 inspired the new moves to make the RIT a low-power dependent technology operable via low voltage solar energy supply.

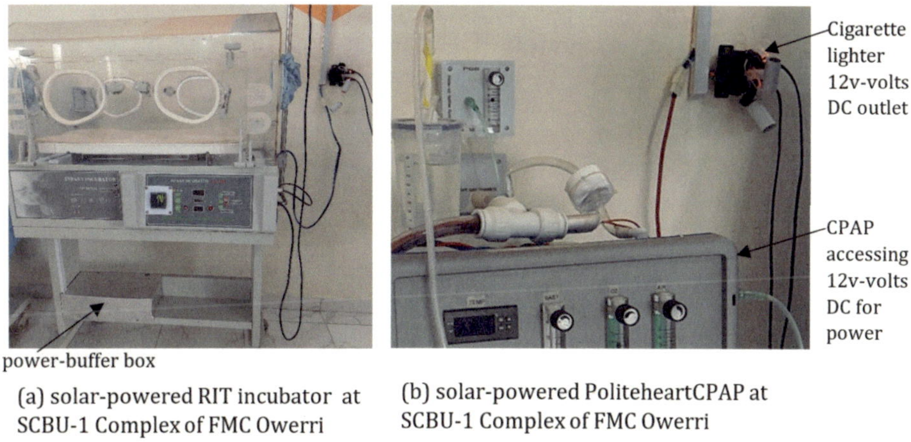

Figure 6.
PLB-powered incubator and CPAP machine (Existing incubator system installed with a power-buffer for running on the low-voltage solar energy and a specially innovated CPAP machine running on real-time solar energy).

2. Invention of the politeheartCPAP v-model (**Figure 6**). A solar-powered neonatal non-invasive ventilator that has contributed to the survival of respiratory-distressed Nigerian neonates since its invention.

3. Innovation of solar-driven conventional phototherapy (**Figure** 7), involving the design and creation of a buffering interface with or without a modified phototherapy light-source to enable existing overhead phototherapy machines such as 'the MTTS Colibri' device to be powered by sunlight using the PLB system

4. Innovation of solar-driven syringe driver (**Figure** 7), involving the design and integration of a solar-buffer circuit to enable systems such as the Graseby 3100 series to run on the solar power driven PLB technology.

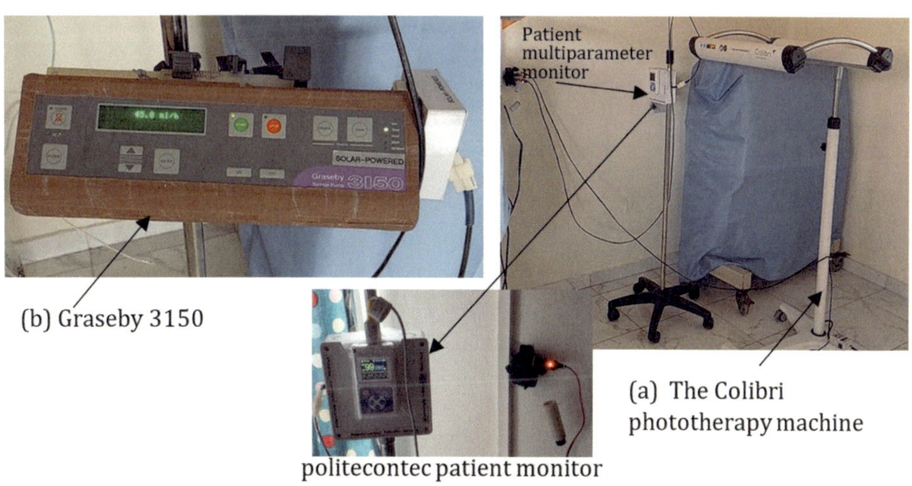

Figure 7.
PLB-powered phototherapy machine, patient monitor, and syringe driver (Device-specific power-buffer circuits have been innovated to allow various existing neonatal devices to run on the low-voltage solar energy, all enabling a near zero-carbon neonatal care delivery).

5. Innovation of the 'politecontec' patient multiparameter monitor (**Figure 7**), involving the design and integration of a solar-buffer circuit to enable a spot-check pulseoximeter to run as a continuous patient monitor on the low-voltage supply of the PLB via a cigarette lighter port.

4.5 The recent invention of the 'politeultralumen (PUL)' phototherapy device

The treatment of severe neonatal jaundice (SNJ) in Nigeria conventionally requires the often-distant travel of the neonate born in faraway places in search of the specialist—it is high risk, labor-intensive, and power-guzzling. Firstly, the mostly inadequate neonatal transport in Nigeria is associated with high morbidity and mortality risks, and secondly, the sought-after specialist gold standard intervention is a surgically invasive high-risk procedure. A non-invasive climate-friendly technique for this condition could minimize these risks. Therefore, it was necessary to turn to this third major concern for neonatal decarbonization.

The treatment of neonatal hyperbilirubinaemia—severe neonatal jaundice—in the newborn in Nigeria is conventionally carried out through an extreme treatment technique called Exchange-Blood-Transfusion (EBT). EBT is very expensive, surgically invasive, and a high-risk procedure. A badly treated hyperbilirubinemia could result in permanent brain damage, poor cognitive abilities, bodily disfiguration, etc.—associated with medical conditions such as kernicterus spectrum disorder (KSD). A lot of LMIC children are behind the doors suffering from complications of KSD. A recent non-invasive treatment technique showed that a well-organized and delivered high-intensity phototherapy irradiation—of the correct light wavelength, could treat over 80% of cases of the diseased blood in-situ, rather than surgically draining the blood out in all cases with its accompanying risks and energy consumption. The non-invasive technique such as the Firefly® phototherapy system (MTTS Asia Co., Ltd., Lane 41 An Duong Vuong, Tay Ho, Hanoi) requires mains high-voltage to operate. This is expensive, largely unaffordable and unusable in the hinterlands where electricity and fossil fuel generators are scarce.

Faced with the challenge of providing a hinterland-operable solution for hyperbilirubinemia in Nigeria, I decided to initiate the development of a climate-friendly device that could apply the principles of high-intensity irradiation to treat severe jaundice in a typical non-electrified Nigerian village. The result was the Politeultralumen device (**Figures 8** and **9**), Nigerian Patent (IPONMW638442007947116728, F/PT/NC/2024/11141).

4.5.1 Zero-carbon effect of the politeultralumen

The device showcases a new idea of using solar-power ONLY to generate high intensity phototherapy irradiation for the treatment of hyperbilirubinemia in the newborn without the need for high-carbon-footprint mains or fossil fuel electric generator. This new non-invasive device has been innovated to target neonatal treatment both in the urban cities and in the hardest-to-reach locations where mains electricity and medical specialist care may not be available. Therefore, this invention delivers power-charging support to appropriate accompanying diagnostic bilirubinometer. The device is operable by local basic trained healthcare workers. It applies a pre-set intensity of irradiation—discretely targeting blood-vessels at the various aspects of interest on the neonate body.

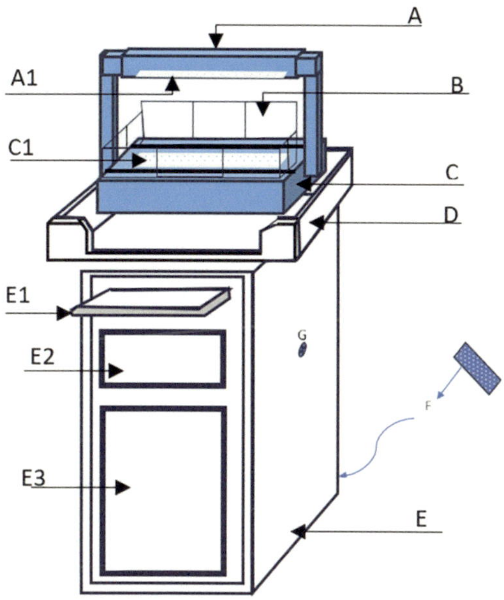

LEGEND

A Top-lamp frame. Made of hollow rectangular tube (7cm x 10cm x 64cm)

A1 Top-light source. 4 x a gang of 32 LEDs delivering 460 nm light wavelength

B Baby bassinette. Manufactured from 4mm gauge clear acrylic Perspex, base dimensions (30cm x 50cm), height: 15cm (front and sides) 20cm (back)

C Base-lamp frame. Made of seven cut-pieces of hollow rectangular plastic tubes and finished to an overall dimension of 34cm x 66cm

C1 Top-light source. 4 x a gang of 32 LEDs delivering 460 nm light wavelength

D System Tray. Manufactured from hardened plastic or ¾ inch water resistant plywood and finished to overall dimension of 42cm x 72cm, and depth of 5cm

E Grounding cabinet. This serves as support and stand for the main irradiation module, and doubling as desk, storage, and powerhouse, height of 75 cm

E1 Retractable worktop for writing and stationing the handheld transcutaneous bilirubinometer

E2 Storage drawer for baby's items

E3 Power unit assembly compartment

F pre-processed energy through solar-panels outside the building

G 12-volt DC, 2.5 Amps outlet port for transcutaneous bilirubinometer charging

Figure 8.
A brief description of politeultralumen drawings (The device is designed to enable the treatment of severe neonatal hyperbilirubinemia at zero-carbon cost to the environment, a compelling shift from the gold-standard combination of exchange-blood-transfusion and electric-phototherapy, known to be a power-guzzling high carbon impact procedure).

4.5.2 Characteristics of the politeultralumen

a. Prior to this innovation, the existing bi-directional phototherapy devices rely on mains electricity to function. However, the politeultralumen can deliver the treatment of hyperbilirubinemia via the rapid breakdown of serum bilirubin in remote places where high-voltage mains electricity is unavailable, or where practitioners would desire to treat the neonate without adding to environmental carbon pollution

b. The innovation features an original idea of an aspect of bi-directional phototherapy powered solely by solar energy based on low-voltage, direct-current, and low-power requirement of light-emitting-diodes (LED)

c. The system is capable of utilizing real-time solar power from sunlight as captured by a gang of solar panels to generate pre-filtered high intensity light of 460 nm wavelength, enabling it to deliver up to 158 $\mu W/cm^2/nm$ irradiation at the couching level of the patient

d. Through the pre-set positioning of LED gangs of its top- and bottom-light sources, the device discretely targets various aspects of neonate's body with the appropriate intensity for a 'total body irradiation' of the patient

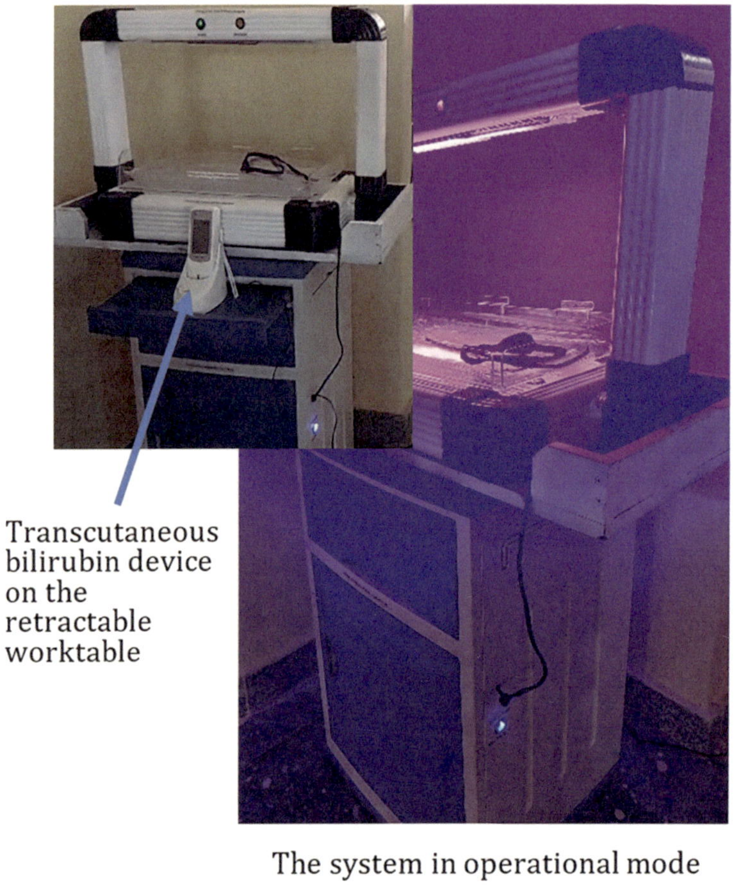

Figure 9.
Manufactured prototype of the 'POLITEULTRALUMEN' device (The politeultralumen device in functional mode, and insert showing its accompanying diagnostic transcutaneous bilirubinometer).

e. The device uses this intensity to rapidly degrade severe jaundice to a lower serum bilirubin level of mild jaundice classification

f. The device uses a total of 100 amp-hour battery storage to privately bank up to 27 hours of full-power operational energy, which it reverts to during sunlight downtimes, thereby freeing it from the questions of how it could carry out its duty in the night

g. The device's baby bassinette is spacious to accommodate typical Nigeria big babies, and the side guards are split at intervals (top-to-bottom) for air cooling, and the far-end side guard is taller to serve as 'wind gust breaker'

5. Conclusions

My power delivery innovations and the tweaked compatible technologies presented in this article have not only broken the barriers for taking neonatology to the majority death-threatened Nigerian (LMIC) neonates in the hinterlands, but these have done so without the usual healthcare delivery damages on the climate and environment. My Nigerian laboratory and mini workshop where the life-size prototypes of these devices are produced—where my staff team assemble regularly for our projects—is powered 100% by solar energy via the PLB and PBS technologies, including all sockets and facility lighting requirements, and all our day and night living power needs for gadgets. This reduces the carbon footprint of my research activities closer to 'zero' since 2018. Our laboratory decarbonization has already been achieved!

It is exciting to note that our ability to use my uniquely combined techniques of the PBS and PLB to generate, bank, and manage 'green energy' for 100% of power requirement to operate and sustain full neonatal healthcare services at the Calabar Women and Children Hospital (CWCH) has, in a classical manner, given the neonates therein-treated, the hypothetical 'Zero-carbon' footprint in the services they receive—the neoroom (neonatology wing of CWCH), which operates on my neonatal technologies and strategies is fully reliant on solar energy as its main source of all required operating power. By successfully taking neonatal traffic to the hinterlands in this manner, my NRS promises to drastically reduce emergency neonatal traffic to the bigger cities, minimizing the associated carbon footprint due to referrals.

The over 15 years of my extraordinary unconventional NRS techniques of solving the Nigerian neonatal problem has been at the core of my Imperial College London research center for frugal medical technology for the LMICs. My core solution pathway—taking our medicine (neonatology) to the hinterlands—where most of the needy neonates are located, was a daunting and almost an impossible journey to embark upon. This required courageous self-driven initiatives of extraordinary unconventional ways of providing energy to do a power-guzzling healthcare delivery. Hence, my burning passion for agenda 'neonatology to the hinterlands to save lives', compelled me to resort to the ever-available solar energy—to modify neonatal technologies for solar compatibility and tweak applications of solar energy storage and usage to make these sufficient for medicine at the hardest-to-reach remote places of the world. My agenda to save neonatal lives has yielded two globally compelling solutions—(1) sustainable healthcare to the remote or isolated people and (2) healthcare delivery without harming the environment. Therefore, the Neoroom of CWCH Calabar, for example, has not only shown how neonates in the hinterlands could receive uninterrupted quality care, but this has also demonstrated the United Nations' dream of 'low- or no-carbon footprint' in a healthcare delivery, and arguably presented Nigeria as a major player in the drive to lower global carbon emission, minimize global warming, and save our planet.

Acknowledgements

I wish to thank all the Nigerian clinicians and nurses who, at various times, supported and played divers roles in these research works. I particularly wish to thank members of my central technical team – Onyeabo Ebenmelu, Jude Nzerem, Chuks Ugbome, and Jerry Okoye. This work was supported by Neonatal Concerns for Africa, www.neonatalconcerns.org. Professor Hippolite Amadi's outreach in Nigeria is supported by the Hornchurch Baptist Church, England, United Kingdom.

Conflict of interest

The author declares no conflict of interest.

Author details

Hippolite Amadi
Centre for Frugal Medical Technology for LMICs, Department of Bioengineering, South Kensington Campus, Imperial College London, London, United Kingdom

*Address all correspondence to: h.amadi@imperial.ac.uk

References

[1] Burgess S. The year 2024 set to end up as the warmest on record. In: Monthly Climate Bulletin of Copernicus Climate Change Service (C3S) 7th November 2024 [Internet]. United Kingdom: ECMWF; 2024. Available from: https://climate.copernicus.eu/year-2024-set-end-warmest-record [Accessed: November 26, 2024]

[2] Garba MJ, Williams EE, Odionye CM, Amadi MA, Godswill JC. A health technology assessment perspective on communicating local knowledge of Hippolite O. Amadi et al.'s neonatal interventions in Global Health research. Asian Journal of Pediatric Research. 2024;**14**(8):36-47. Available from: https://journalajpr.com/index.php/AJPR/article/view/379

[3] Amadi HO, Olateju EK, Kawuwa MB, Osibogun AO, Alabi P, Ibadin MO. Neonatal hyperthermia and thermal stress in low- and middle-income countries: A hidden cause of death in extremely low-birthweight neonates. Paediatrics and International Child Health. 2015;**35**(3):273-281. DOI: 10.1179/2046905515Y.0000000030

[4] Amadi HO, Mokuolu AO, Obasa T. The effect of high sunlight intensity on the neonatal incubator functionality. Journal of Neonatal Nursing. 2012;**19**(3):122-128. DOI: 10.1016/j.jnn.2012.03.011

[5] Amadi HO. Neonatal thermoneutrality in a tropical climate. In: Rodriguez-Morales AJ, editor. Current Topics in Tropical Medicine. London, UK, Croatia: IntechOpen Access Publishers; 2012. pp. 513-544. Available from: http://www.intechopen.com/books/current-topics-in-tropical-medicine/neonatal-thermoneutrality [Accessed: March 16, 2012]. ISBN 978-953-51-0274-8

[6] Amadi HO, Kawuwa MB, Mohammed LI, Mohammed H, Oyedokun A. Eradicating Climate Induced Neonatal Hyperthermia through Nursery Building Design. Second Global Forum on Medical Devices. Geneva, Switzerland. Available from: http://apps.who.int/medical_devices/2ndWHOGlobalForum-OralPresentations271113.pdf: World Health Organisation; 2013 [Accessed: February 15, 2014]

[7] Woods J, James N, Kozubal E, Bonnema E, Brief K, Voeller L, et al. Humidity's impact on greenhouse gas emissions from air conditioning. Joule. 2022;**6**:726-741. DOI: 10.1016/j.joule.2022.02.013

[8] Amadi HO, Mohammed IL, Kawuwa MB, Oyedokun A, Mohammed H. Synthesis and validation of a weatherproof nursery design that eliminates tropical evening-fever-syndrome in neonates. International Journal of Paediatrics. 2014;**2014**:9. DOI: 10.1155/2014/986760. ID 986760

[9] Amadi HO, Kawuwa MB, Abubakar AL, Adesina CT, Olateju EK. A community integrated concept that minimises death of most vulnerable neonates at poor-resource environments. Journal of Paediatrics and Neonatal Care. 2022;**12**(3):170-173. DOI: 10.15406/jpnc.2022.12.00475

[10] TVC News. Yobe expands primary, secondary, and tertiary health facilities – adopting a systemic approach to ensure healthcare accessibility. 2023. Available from: https://www.youtube.com/watch?v=istvgePNWCY

[11] CNN International. Energy Innovators in East and West Africa: In

inside Africa Show. London, United Kingdom: Turner Broadcasting System Europe Limited; 2022. Available from: https://edition.cnn.com/videos/tv/2022/07/04/energy-innovators-nigeria-ghana-malawi-spc-intl.cnn

[12] Amadi HO, Kawuwa MB, Abubakar AL, Obaro SK. Fundamentals of a safe and effective neonatal building design in a tropical LMIC setting. In: Mauricio BR, editor. Best Practices in Neonatal Care and Safety. Croatia: IntechOpen Access Publishers; 2023. DOI: 10.5772/intechopen.1002399. Available from: https://www.intechopen.com/online-first/1155918. ISBN: 978-1-78923-363-6http://hdl.handle.net/10044/1/106350 [Accessed: September 19, 2023]

[13] Amadi HO, Abubakar AL. LMIC facility-lighting limitation in Nigeria fully resolved by a novel frugal polite-light-bank technology. Global Journal of Medical Research. 2023;**GJMR-K 23**(3):1-6. Version 1.0

[14] Amadi HO, Azubuike JC, Etawo US, Offiong UR, Ezeaka C, Eyinade O, et al. The impact of recycled neonatal incubators in Nigeria: A 6-year follow-up study. International Journal of Paediatrics. 2010;**2010**:7. Article ID 269293. DOI: 10.1155/2010/269293

[15] Amadi HO, Mokuolu O, Adimora GN, Pam SD, Etawo US, Ohadugha CO, et al. Digitally recycled incubators: Better economic alternatives to modern systems in low-income countries. Annals Tropical Paediatrics. 2007;**27**:207-214

Chapter 2

Perspective Chapter: Oil Palm Plantations Can Offset Carbon Loss and Improve Livelihoods of Rural Communities in Africa

Paul L. Woomer and Mpoko Bokanga

Abstract

The oil palm is endogenous to the humid tropical belt of West and Central Africa. Its cultivation was greatly expanded in Southeast Asia, and today, it accounts for 85% of commercially planted oil palm in the world. Oil palm plantations in Africa could become eligible for accrued carbon credits under some strict conditions and contribute to achieving sustainable development goals in Africa. Plantations must not be recently carved from humid forests nor established on peat soils, as the comparative long-term carbon stocks remain unfavorable. However, longstanding plantations or those established on mineral soils of grassland and degraded cropland offer a strong potential to accumulate system carbon over decades. The upper limits of this accrual are manageable and reliable methods to monitor system carbon gains are available. Carbon emissions may also be reduced through improved management of the plantations' palm oil mills, and through conversion of waste plantation biomass to biochar products. The revenues generated from plantation carbon offset payments should be directed toward improving the livelihoods of rural communities established around the plantations and toward the protection of adjacent natural wildlife habitats. An example of how plantations in DR Congo could qualify for and implement a carbon credit program is provided.

Keywords: Africa, biochar, carbon sequestration, corporate social responsibility, Democratic Republic of Congo, *Elaeis guineensis* Jacq., humid forest zone, International Institute of Tropical Agriculture, private sector engagement

1. Introduction

Global efforts are underway to establish CO_2 mitigation policy amid uncertainty about the deeper economic consequences of its actions. Common wisdom currently holds that palm oil producers should not be considered as viable candidates for carbon offset credits because their plantations historically result from the conversion of tropical forests. This position may not be tenable, particularly in Africa, where plantations were established many decades ago and now drive local economies. Bokanga [1] describes the importance of oil palm (*Elaeis guineensis* Jacq.) in the Democratic Republic of Congo (DRC) and argues that, if responsibly developed, the palm oil industry could lift millions of people out of poverty. With its plentiful humid land, abundant labor, favorable climate and reliable market demand, the DRC is poised to become the next frontier for palm oil, but in a way that learns from past mistakes made in Asia and elsewhere, ensuring human prosperity, social progress and environmental protection. An important component of this future is the award of carbon credits for its carefully informed efforts and then the use of these credits for improvements in the social welfare of communities living within oil palm-forest ecosystems.

Planned development of oil palm expansion in South America provides a useful example in terms of sustainable development. Over the past decade, new plantations have increasingly avoided deforestation and are guided by roundtable certification programs [2, 3]. Strategies are also being adopted to expand oil palm coverage in Indonesia and Malaysia in ways that avoid deforestation and peat soils [4, 5]. Substantial growth of the palm oil industry is expected in Africa and there is commitment to achieve such growth in a manner that improves livelihoods and protects natural resources, including forests [1]. Numerous countries of West and Central Africa signed the Marrakech Declaration for Sustainable Development of the Palm Oil Sector in 2016 during COP-22 in which they commit to sustainable development of the palm oil value chain. This effort was reinforced through the African Palm Oil Initiative, followed by the Africa Sustainable Commodities Initiative, where these countries commit to zero deforestation [6]. Within the DRC alone, the Ministry of Agriculture identified over 2000 abandoned farms, many of which can be put into palm oil production at no risk to primary forests [7]. More expansive goals toward a sustainable palm oil industry are now directed through participation in the Roundtable on Sustainable Palm Oil (RSPO) [8, 9].

The main environmental concern related to the palm oil industry is deforestation and its effect on habitat loss and the release of system carbon into the atmosphere. Secondary impacts include pollution of local surface waters and excessive use of mineral fertilizers and pesticides. In response, the industry established the RSPO certification program [8] to ensure that palm oil has a positive impact on the planet and on people. Stakeholders engaged in this effort include oil palm producers, processors and traders, as well as manufacturers, investors, environmental conservation groups and an assortment of Non-Governmental Organizations. It is notable that oil palm plantations and mill managers are increasingly adopting practices that reduce negative environmental impacts [10] through site-specific nutrient balance management, biological control of pests and establishment of leguminous cover crops, particularly *Mucuna* and *Pueraria* spp. [11].

When evaluating the importance of oil palm, we must consider that this tree crop yields 5–10 times more oil per hectare per year than annual oil crops. In this way, oil palm occupies less than 10% of the land planted with oil crops but produces more than 35% of the oil consumed worldwide [12]. Oil palm requires less land, pesticides and fertilizers and provides income to some of the world's poorest countries. Yet oil palm remains criticized for its negative impact on GHG emissions in ways that tend to exclude it from the award of carbon credits. We contend that this situation requires reevaluation.

2. The controversy surrounding oil palm

Oil palm plantations have a very poor reputation, but much of this disdain is likely undeserved. Oil palm plantations are viewed as major drivers of tropical deforestation that leads to massive emissions of C into the atmosphere. Moreover, it has been estimated that, given the current carbon credit system, converting forests to palm oil production was much more profitable than conserving forests for carbon credits [13], which could potentially drive additional rounds of forest loss. Germer and Sauerborn [14] estimated the impact of oil palm plantation establishment on system C balances through both forest and grassland conversion. Forest conversion on mineral soils resulted in losses of 177 t C per ha and considerably more in peat soil. In contrast, conversion of tropical grassland to oil palm resulted in 36 t C ha^{-1} gains in system C. Hashim et al. [15] confirmed the vulnerability of C loss from peat soils converted to oil palm production, estimating losses as great as 77 t CO_2 eq ha^{-1} depending upon the depth of the peat deposit and the amount of nitrogen fertilizer applied over time. In the Kalimantan region of Indonesia, oil palm expanded by 278% from 2000 to 2010, with 90% of these lands derived from intact forests (47%), logged areas (22%) and agroforests (21%). Ominously, 79% of Indonesia's allocated leases for conversion to oil palm remained undeveloped as of 2010 [16]. The full development of these leases, which cover 93,844 km^2 (90% forested lands, including 31% intact forests), would contribute 18–20% (0.12–0.15 GtC yr^{-1}) of Indonesia's 2020 CO_2-equivalent emissions.

Expansion of oil palm cultivation into African humid forests could lead to large carbon emissions. In Gabon, the conversion of 11,500 ha of logged forest into oil palm plantations is expected to release about 1.50 Tg C. However, Burton et al. [17] have estimated that these emissions could be offset over 25 years through sequestration in the planned forest set-asides given a 2.6:1 ratio of logged to converted forest. It is important that African land use planners have means to reduce these expected environmental impacts through a better understanding of system C stocks and incentives offered for their better management. Using original forests as a reference point, numerous ecosystem functions are lost in oil palm plantations including those with global impacts such as carbon storage, habitat loss and incompletely described and preserved genetic and medical resources [18]. One function that increases greatly within palm plantations is the generation of economic value. Fitzherbert et al. [19] assert that oil palm plantations support much fewer species than do forests, and less than other tropical tree crops, concluding that substantial biodiversity losses can only be averted by assuring future oil palm expansion avoids direct deforestation [20].

3. Oil palm carbon stocks and sequestration

Forest and oil palm biomass were compared in Indonesia by Kotowska et al. [21]. Total tree biomass in natural forests averaged 384 Mg ha^{-1} and was over seven times greater than average oil palm plantation biomass (50 Mg ha^{-1}). However, Net Primary Productivity was higher in the oil palm system (33 Mg ha^{-1} yr^{-1}) than in the natural forest (24 Mg ha^{-1} yr^{-1}), in large part because of massive offtake from fruit production (15–20 Mg ha^{-1} yr^{-1}). Oil palms compete less with one another and receive the benefits of soil and pest management, and in this way, their potential to sequester C in the short term may be greater than that of forests. Total system carbon in an oil palm plantation in eastern Amazonia was estimated to hold 99.1 ± 3.1 Mg C per ha including 2.3 ± 0.1 Mg C ha^{-1} of litter [22]. The total dry biomass in another 25-year-old Brazilian plantation contained 90 Mg ha^{-1} [23]. An oil palm plantation accumulated 112 Mg ha^{-1} aboveground biomass that sequestered 3.7 Mg C ha^{-1} yr^{-1} after 10 years in northeast India [24]. Citing many sources from Indonesia, the Palm Oil

Agribusiness Strategic Policy Institute [25] suggests that the carbon stocks of oil palm plantations range from 30 Mg C to 75 Mg C ha^{-1}, with a mean of about 40 Mg C ha^{-1} in an average-aged plantation. From these examples, a strong pattern is emerging across diverse locations.

A detailed C budget of oil palm involving plant biomass, litter and soil, also reported high rate of carbon sequestration by oil palm ecosystems, between 2.5 and 9.4 Mg ha^{-1} yr^{-1}, including harvested bunches [26]. Similar findings were reported by Lewis et al. [27] who recorded aboveground biomass accumulation in Malaysia of 6.4 Mg ha^{-1} yr^{-1} during the first 12 years after planting, and 8.0 Mg ha^{-1} yr^{-1} thereafter. Research from the Philippines reports that a 9-year old oil palm plantation sequestered 6.1 Mg C ha^{-1} yr^{-1}, with total C stocks of 55 Mg C ha^{-1} [28]. Note how these findings fall into general agreement, particularly the rates of annual sequestration. Clearly, the oil palm has major potential for atmospheric CO_2 sequestration over the short term, a factor that must be better appreciated in judging its larger environmental impacts.

Oil palm plantations should be eligible for accrued carbon credits under some strict conditions. Carbon debt results when the vegetation removed to establish crops have a greater C stock than the resulting managed system. In a study of 25 oil palm plantations in Indonesia, the average aboveground C stock of a 25-year rotation was 42 Mg C ha^{-1} on mineral soils and 40 Mg C ha^{-1} on peat soils. In contrast, in Chiapas, Mexico, a 12-year-old oil palm plantation accumulated 878 kg biomass C per palm, or 126 t C ha^{-1} in a stand of 143 palms ha^{-1} [29], suggesting that large variability exists between plantations and areas. Henson et al. [30, 31] stated that where lands converted to oil palm have reduced biomass, plantations can lead to increased carbon stocks. In this case, emissions from land use change are reduced compared to other sources of greenhouse gasses from mills, fossil fuels and fertilizers. From this, we conclude that plantations must not be recently carved from humid forests as the comparative carbon stocks remain unfavorable into the foreseeable future. Plantations also must not be established on peat soils as the long-term dynamic of soil organic matter is strongly negative and complicates carbon accounting. However, longstanding plantations or those established on the mineral soils of grassland and landscapes degraded by previous human activities offer a strong potential to accumulate system carbon over decades.

Henson et al. [32] described the importance of including frond bases within estimates of palm biomass and system carbon stocks of the oil palm. These frond bases remain attached to the trunk after frond pruning. Estimates of C stocks in Papua New Guinea were increased by 11% after taking frond bases into account. Soil C increases were documented in Brazilian oil palm plantations [33]. Soil C stocks were higher adjacent to the oil palm base and beneath frond piles compared to access pathways. In some cases, a soil organic C increase of 25% was observed compared to soils under native vegetation, indicating a substantial gain of soil C stocks in oil palm plantations over time.

4. Better managing oil palm plantation carbon

Reliable and efficient methods to monitor system carbon gains are available. Oil palm biomass is primarily a product of its' combined root, trunk and frond biomass, with an additional value assigned to its fruit removal [34]. Determining the capacity of oil palm to sequester carbon requires accurate biomass estimates based upon destructive sampling that is used to develop and test allometric equations, as was performed in Gabon by Migolet et al. [35]. Stems accounted for 73% of total aboveground biomass, rachises for 13%, petioles for 8% and fruits and leaflets for 6%. Roots are assumed to hold about 30% of the total biomass. The best allometric equations incorporate Diameter at Breast Height (DBH), stem height, tissue density and leaf number, although the latter two parameters are more difficult to measure. A simpler model relying upon DBH and height accounted for 93% of observed aboveground biomass.

Established relationships also exist between canopy size and aboveground biomass, with both crown diameter and crown area serving as proxy variables [36]. An even simpler allometric equation developed in Mexico considers only oil palm height using the equation Biomass = 98.349 Height + 737.41 (R2 = 0.577) [37]. Allometric equations are also available to calculate the average dry weight of mature fronds as a means of estimating annual production of oil palm frond biomass and their contribution to plantation system carbon stocks [38]. While it is not the purpose of this paper to compare different allometric approaches and equations, it is important to note that such tested equations are widely available in the published literature.

Several plantation management practices lead to increased carbon stocks within oil palm plantations. The upper limits of this accrual are manageable based on the handling of crop residues and accompanying vegetation and basic soil conservation measures. Plantation management measures prevent or reduce the loss of ecosystem functions by avoiding further land clearing and burning, avoiding peat soils, relying upon integrated pest management and introducing of cover crops, mulching and composting [18]. Carbon-smart management of oil palm residues suggests that they are best returned as mulch for more efficient C and nutrient cycling. Leaf biomass pruned from oil palm and applied as mulch steadily decomposes to provide a source of organic matter and nutrients. During 2 years of decomposition, fronds and leaflets lost 88% and 86% of their initial weight and released 51% and 83% of their nitrogen, while also releasing 87% and 93% of phosphorus, respectively [39]. Oil palm readily recycles its nutrients.

Residue retention as a part of management practice allowed for increased soil organic carbon stocks within oil palm plantations in Sumatra, Indonesia. Soil carbon stocks increased by 11 t C ha^{-1} when alternate interrows received enriched mulch and fronds compared to the adjacent unmulched harvesting path [40]. This adjustment allows for access to the fruit bunches while more effectively recycling organic residues, but it must be balanced with the advantages of field sanitation and safety. Indeed, sequestration of soil carbon in these plantations can be a key component to reverse system carbon losses.

Guidelines were established to reduce greenhouse gas emissions from oil palm plantations in Colombia. Suggested practices include increasing the use of organic fertilizers, planting oil palms on degraded land, using oil palm biodiesel as a substitute for fossil fuels, improving the yield of oil palm fresh fruit bunches through increased production efficiency and producing and applying biochar at the time of replanting [41]. It was also calculated that each ton of oil palm fresh fruit bunches resulted in the fixation of 606 kg of CO_2.

5. Alternate pathways for mitigation

There is also the possibility to reduce carbon emissions through improved management of the palm oil mills associated with these plantations. Palm oil processing also leads to greenhouse gas emissions. Hong [42] suggests that a typical mill produces emissions of 637 to 1131 kg CO_2 eq per ton of crude palm oil (CPO) produced and the need exists to reduce this footprint. Capturing biogas from mill effluent, converting palm sludge oil into biodiesel and briquetting biomass to serve as a substitute for fuelwood can reduce these emissions by 457 kg CO_2 eq per ton of CPO. Efforts to redesign mill operations in Thailand demonstrate how the integrated utilization of oil palm biomass has great potential to reduce carbon emissions, with savings of 812 kg CO_2 eq per ton of CPO [43, 44]. In Colombia, 58% of mill emissions originate from anaerobic ponds used for water treatment, 41% from powering mill operations and only 1% from transportation and heavy machinery [45], and corrective measures focus on those first two sources. Indeed, capturing emissions from mill effluent and reducing dependence upon fossil fuels are the two prominent mechanisms for GHG reduction within milling operations [46].

Durable carbon removal of oil palm waste can also be achieved through biochar production *via* thermochemical conversion under no or limited oxygen. The fixed carbon content of biochar and properties relevant for soil health improvement depend on the type of oil palm waste, temperature profile and airflow during processing [47]. Fronds converted at 500°C over 30 minutes produced 40% porous biochar with a calorific value of 22.6 MJ kg^{-1} [48]. Empty fruit bunches offer a particularly suitable candidate for biochar, as shown in SE Asia [49]. There are two major processes for biochar production, often confused but critically different: pyrolysis and gasification. Pyrolysis occurs in the absence of oxygen, relying on external heat from a combustion furnace. This process typically operates at temperatures between 300°C and 600°C. Gasification admits small amounts of oxygen and does not require external combustion. This process usually operates at temperatures ranging from 450°C to 1500°C. Attention must be paid to the differences between the processes since they lead to variable outcomes in carbon balance and biochar characteristics.

A wide range of technologies exists to produce biochar that serves various energy applications. These include domestic cookstoves, simple kilns, rapid food dryers and electricity and heat generators. Artisanal or automated kiln systems can be employed for local in-field production, including continuous flow containerized and cottage-style setups that capture heat for value-addition processes. Recovery of biochar from biomass input for these systems is typically 25–30%, and output capacity ranges from 200 kg to 2.4 ton biochar per batch. Existing gasification setups are self-contained and convert up to 1 ton of dry biomass feedstock into 1 megawatt hour (MWh) of electricity, 1.2 MWh of heat and 100–150 kg of biochar. Each MWh of electricity produced by gasification can replace 400 liters of diesel. Overall, pyrolysis and gasification have fundamental advantages over biomass combustion as an energy source by realizing higher energy efficiency and the ability to operate fully upon low-grade residues obtained from oil palm plantations.

The use of locally available biomass for biochar and the potential for remote production can stimulate rural economies and create jobs. In Malaysia, pyrolysis of oil palm waste has attained commercialization and gasification is being piloted [50]. In comparison, agribusinesses are implementing similar operations within coffee mills in Brazil and rice mills in Thailand. Unit costs of biochar production depend upon logistics, energy utilization and available carbon credits. Costs are lowest when biomass is sourced within less than 10–20 kilometers and when the project is accredited for high-quality carbon credits with minimal administrative overheads for monitoring, reporting and verification.

While biochar may be pressed into briquettes for use as solid fuel, it is often better utilized in other applications. These include serving as an absorbent in chemical processes, a conditioner in compost or manure treatment, a soil amendment in plantations and a potting substrate in nurseries. Harson et al. [51] examined biochar production from palm oil empty fruit bunches in Malaysia. The energy yield from slow pyrolysis was positive, resulting in about 11 MJ kg^{-1} of product, calculating that biochar production costs were \$533 t^{-1}. When subjected to microwave pyrolysis, oil palm wastes produced a biochar with many desirable properties that may be used as a biofuel with high heating value (23–26 MJ/kg), an environmentally friendly adsorbent of contaminants, and as a carbon sequestering soil amendment and biofertilizer [52]. Razali and Kamarulzaman [53] contend that biochar produced from the palm oil trunk is suitable for the absorption of pollutants within mill wastewater effluent. These results suggest that oil palm waste is better transformed into biochar than simply disposed or burned as a low-grade fuel.

Kong et al. [54] reviewed the potential and challenges of producing biochar from oil palm biomass, concluding that biochar leads to a healthier environment and societal and economic growth for the oil palm industry, as well as benefiting global sustainability interests. Oil palm seedlings raised on potting mixture with 3% w/w biochar from empty fruit bunches resulted in 23% faster growth in the first year compared to when fertilizer alone is used [55]. Long-term research trials in annual cropping systems of Kenya found decade-long increases in yield and fertilizer use efficiency from a single biochar application [56]. The use of biochar is becoming commonplace in coffee agribusiness across the Global South, with producers in Brazil and Tanzania applying it since 2011.

When produced under the right conditions, biochar removes carbon from the atmosphere for hundreds of years [57]. In 2023, biochar was leading the delivery of carbon removal among all strategies, favored by credit markets due to affordability, permanence, ease of certification, scalability and benefits to communities. Robust chain-of-custody application is available for artisanal production that tracks all steps of the process from sourcing to conversion and sinking. Projects with combined biochar and energy production may need to invest in the development of integrated data management systems for streamlining the quantification of offsets and removals at various levels of operation. Biochar is currently not yet recognized as a soil input by the government of the Democratic Republic of Congo (DRC). However, its use has been recommended by local experts [58] and has been confirmed as effective in improving water retention in soils [59]. Various international organizations have established quality standards for biochar, focusing on aspects such as carbon content, nutrient levels and limits on heavy metals and organic contaminants. These standards differ based on the intended application of biochar, whether as a soil amendment, energy source, or building material.

6. Carbon credit qualification and associated risks

Any carbon credits awarded to palm oil producers in Africa should be in full compliance with the principles and criteria of the Roundtable on Sustainable Palm Oil [8, 9]. RSPO is a multi-stakeholder organization that guides the palm oil industry toward environmentally sound and socially equitable outcomes. It offers comprehensive standards and procedures that support climate security. RSPO requires members to understand system carbon stocks and prohibits planting in forests and on peat soils. It also works with members to calculate emissions risk profiles of processing facilities. It offers a reasonable balance between environmental protection and economic development and estimates that its guidance avoids the release of 2.2 million MT of CO_2 eq yr^{-1} [9].

Before credits for carbon removal through palm oil waste management are sold, the supplying entity must demonstrate compliance with criteria on baselines, additionality, quantification and monitoring, permanence and avoidance of double accounting. These are governed through protocols from carbon registries that are internationally accredited for carbon offset projects to document and validate project design and monitor periodic reports. Independent rating agencies can also score projects based upon various factors related to planned environmental and social impacts. In this way, levels of supervision also affect the assigned price per unit of GHG removal.

There may be issues with additionality, as projects must demonstrate that carbon removal would not have taken place without key policy and tax incentives, or without credit revenue. Additionality is determined by assessing whether the proposed project is distinct from its baseline scenario. This may form another challenge as projects must have historical data on land expansion and tree cover loss, but its requirement forces plantations to better monitor their environmental impacts over time.

7. The case for PHC as an eligible C offset provider

Plantations et Huileries du Congo (PHC) is the largest industrial producer of palm oil in the Democratic Republic of the Congo (DRC). PHC's headquarters are located in Kinshasa, but its plantations are located deep within the Congo Basin at Lokutu (Tshopo Province), Yaligimba (Mongala Province) and Boteka (Equateur Province). The company manages over 100,000 hectares of land within its concession. Oil palm occupies 28% of this concession, while 24% is adjacent forests of high conservation value. Fallows and degraded forests cover 46% of the concession, while mills, offices, housing and roads occupy the remaining 2%. Established in the 1910s, Boteka is PHC's smallest and oldest plantation. The Yaligimba and Lokutu plantations were established in the 1920s and 1930s, respectively. Each plantation hosts a palm oil mill that is powered by a turbine, obtaining energy from a boiler that burns biomass from plantation harvests. The boiler also produces steam that is used during the processing of oil palm's fresh fruit bunches. The mesocarp of the oil palm fruit provides crude palm oil (CPO). In a separate process, the kernel of the oil palm fruit is obtained and cracked to release its nut, which is pressed to extract a second type of oil, the palm kernel oil (PKO). No chemicals are used in the production of palm oil. In 2023, PHC managed mature oil palms covering about 21,400 hectares.

A projection based upon data from this paper and information on production and mill practices suggests that the 21,400 ha under oil palm production sequester 76,397 ± 3777 tons of system carbon per ha per year. This is equal to 280,124 ± 13,847 tons of CO_2 eq that could be worth US $1,400,618 ± 69,236 per year, at $5 per ton CO_2 eq, if a buyer could be found (**Table 1**). This projection assumes a 25-year rotation starting with 140 palms per ha and an attrition of 0.5 plant ha^{-1} yr^{-1}. It also assumes an equal division of age classes over the 25-year rotation, where 856 ha of land is replanted each year. A projection of plantation carbon over a 25-year rotation appears in **Figure 1**. This estimate does not consider soil organic C, does not separate leaf and fruit C over time and requires further verification within each plantation in the future.

It is likely that the greatest environmental benefits are obtained as the result of producing biochar from the oil palm's empty fruit bunches and then returning the

Parameter	Total C gain	CO_2e gain	Offset value
	Mg yr^{-1}	Mg yr^{-1}	US$ yr^{-1}
Scenario 1	67,148	246,209	1,231,047
Scenario 2	81,152	297,557	1,487,787
Scenario 3	80,892	296,604	1,483,020
Mean	76,397	280,124	1,400,618
SEM	3777	13,847	69,236

Table 1.
Estimated C gains across the PHC plantations, their CO_2 equivalents and offset values at \$20 t CO_2e based on three scenarios drawn from the literature cited in this paper (SEM = Standard Error of the Mean).

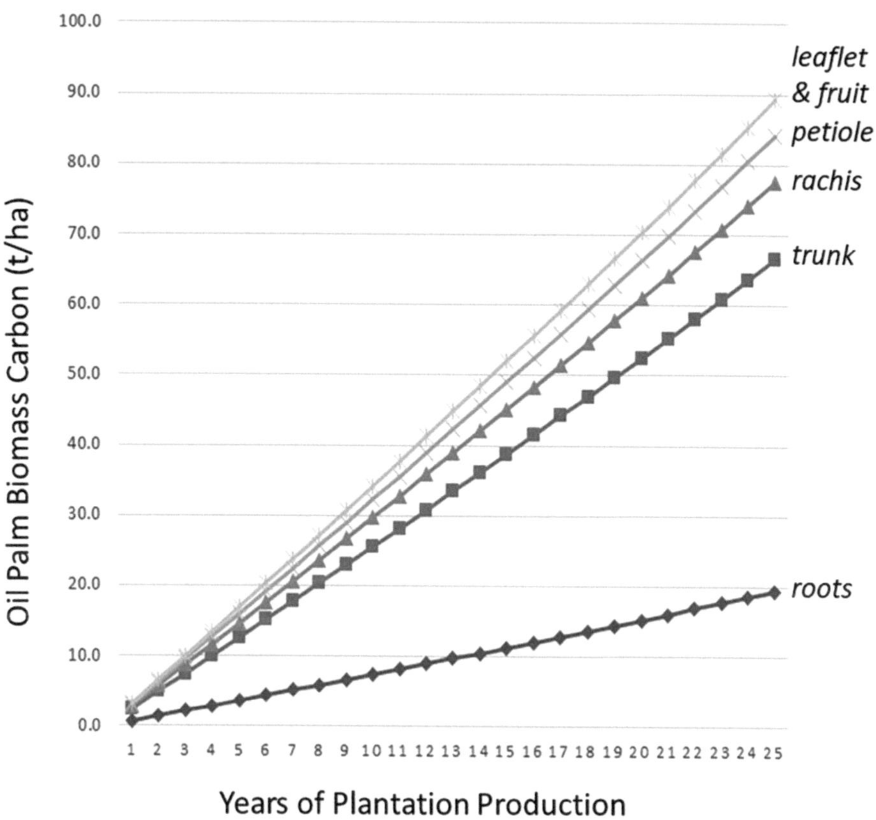

Figure 1.
Estimated C accumulation in the PHC oil palm plantations over their 25-year rotational cycle.

product to the plantation's soils. PHC collects about 70,000 tons of these empty fruit bunches from its mills per year (**Table 2**) and transports them back to the plantation. Converting them to biochar at 17.5% biochar yield, the equivalent of 12,250 tons of biochar could be obtained. Applying this biochar to annually replanted lands (858 ha yr^{-1}) would contribute 14.3 tons of biochar per ha, or 102 kg of biochar per palm. If the same quantity of biochar were to be spread across the entire plantation, each ha would receive 0.57 t yr^{-1}, equivalent to 4.4 kg of biochar per palm. It is not yet determined

Parameter	Amount
Empty fruit bunches (EFB) collected per year (t yr^{-1})	70,000
Biochar C produced from EFB (t yr^{-1} at 17.5% yield)	12,250
Lands replanted per year (ha)	856
Total seedlings replanted per year (no)	119,840
Biochar applied to replanted area only (t ha^{-1})	14.3
Biochar applied to replanted seedling only (kg $plant^{-1}$)	102
Biochar distributed across entire plantation (t ha^{-1})	0.57
Biochar distributed across all oil palms (kg $plant^{-1}$)	4.3
Value of applied biochar at $5 per t CO_2e ($ yr^{-1})	$61,250

Table 2.
Benefits from producing biochar from oil palm's empty fruit bunches and then sinking that carbon into plantation soils.

which scheme would be most practical and profitable, but evidence suggests that biochar addition also improves the use efficiency of applied mineral fertilizers.

The paper also explores how a carbon credit program for the three extensive plantations of PHC Ltd. in DRC could be structured. The revenues generated from plantation offset payments are best directed toward the communities living around plantation activities and toward the protection of the adjacent natural habitat. Climate financing directs funds in support of mitigating climate change and requires that land managers adopt a suite of monitored practices aimed at supporting efforts to mitigate and adapt to climate change. It involves financial instruments such as grants, concessional loans, equity investments and guarantees. When used as a financing mechanism for carbon management by large, tropical plantations, it catalyzes innovation that advances more sustainable agricultural practices and technologies that in turn sequester additional, more predictable quantities of system C into the future. In the case of PHC and its plantations in the Congo Basin, this financing would promote climate-smart plantation practices by the company and climate-smart agricultural production systems to be adopted by surrounding communities, facilitate data collection and analysis of oil palm systems and adjacent natural forests, foster markets and payments for expanded ecosystem services and more fully align DRC with the priorities and criteria of climate finance institutions as a means of achieving its climate agreement targets. Several risks associated with carbon offsets involving palm oil producers and possible responses appear in **Table 3**.

Awarding C credits to PHC can also protect the biodiversity of the forests within and adjacent to its concessions. Presently, the loss of natural forest is not occurring as a direct result of plantation practices, but forest disturbance is caused by the slash-and-burn farming practices of local communities in their quest to produce food crops. In addition, illegal logging of *Pericopsis elata*, an endangered species and *Afzelia africana*, a vulnerable species, is ongoing [60]. Swamp forests are common in the area and two of the concessions (Boteka and Lokutu) have highly fragmented plantations because oil palm is not suited to their waterlogged soils. The resulting fragmentation requires that wildlife corridors be established and maintained. At the same time, there are several invasive plant species present in and around the plantations including *Canna indica*, *Mimosa pigra*, *Chromolaena oderata* and *Alchornia cordifolia* [60]. Funds from carbon credits could be directed in part toward their removal. A plan for the

Risk 1. Palm oil producers expand operations into adjacent forests or peat soils despite agreements to the contrary. *Response:* Licenses to start a new oil palm plantation require providing the exact location of oil palm planting and acceptance of periodic monitoring. Plantations are monitored through remote sensing, any infringement upon forests results in disqualification of carbon credits.
Risk 2. Palm oil producers are unable to discourage adjacent communities from disturbing common forest margins. *Response:* Forest infringement is red flagged, local communities are informed about the threat of slash-and-burn practice and alternatives to slash-and-burn agriculture are offered. Access to markets for non-timber forest resources is promoted.
Risk 3. Carbon sequestration within optimally managed plantations does not achieve projected offset levels. *Response:* Payments will be based upon documented carbon gains up to an established limit; the agreement will specify how and how often these gains are measured and offset payments will be made accordingly.
Risk 4. Oil palm plantations developed upon degraded lands and poorer soils are underproductive. *Response:* Plantations are allowed to expand into disturbed lands within their concessions, including previously disturbed forests and derived grasslands, and payments are based upon measured, not projected, carbon sequestered within those lands.
Risk 5. Communities promised a share of carbon credit revenues remain dissatisfied with opportunities for economic improvement. *Response:* An established fraction of carbon payments is earmarked for communities in return for their cooperation within the agreement; local committees will help determine how these funds are allocated with attention to preventing capture by local elites.
Risk 6. Biochar production intended for sinking C into plantation soils proves too expensive or complicated. *Response:* Sinking biochar into soils is an established mechanism of carbon sequestration and plantations have the right to determine how and where these additions occur; biochar may also be used as an energy source to reduce dependence upon fossil fuels or firewood.
Risk 7. Despite strong evidence and arguments to the contrary, carbon investors remain biased against palm oil producers. *Response:* The credibility of carbon sinks is left to those willing to pay for them, but when African countries include the management of plantation carbon in their national plants, these become more legitimized.

Table 3.
Possible risks associated with C offsets among palm oil producers in Africa and best possible mitigation responses.

control of illegal logging would also be put into effect. Finally, PHC's ongoing community development initiatives will increase household food supply and reduce the need to clear additional forests.

The southern bank of the Congo River, where the Boteka and Lokutu concessions operate, is also the habitat of the great ape *Pan paniscus*, better known under the common name "bonobo." This endangered species is the closest relative of the human species and is only found in this part of the DRC [61]. Its survival is threatened by poachers and the slash-and-burn agriculture that disturbs its habitat. PHC provides support to a project for the protection of the bonobo to prevent its hunting, to offer alternatives to slash-and-burn agriculture and to provide sustainable livelihoods to the indigenous communities that share the bonobo habitat. Resources from carbon credits will strengthen these efforts and provide incentives for additional conservation measures, including through mobilizing supportive community actions.

To its credit, PHC has pledged to promote environmental management systems that are sustainable, minimize environmental degradation and protect forests and habitats along its margins. This approach offers a win-win strategy as it also reduces production costs and increases profitability. Previous plantation practices may have contributed to GHG emissions but that was decades before this impact was widely understood. Sustainable plantation management also avoids soil erosion, nutrient depletion and pollution of surface and groundwater. These approaches to plantation management mitigate climate change by sequestering carbon in the biomass and soil but also improve the resilience of plantations themselves, making them more sustainable into the future. These plantations also generate their own bioenergy to operate mills and power surrounding communities.

In addition, PHC is committed to improving living conditions in rural areas [62]. Through its 4 hospitals and 460 patient beds, PHC provides medical care to more than 150,000 people, including its employees and their families and members of the communities surrounding its operational sites. Between 2021 and 2024, the company has built and equipped 24 primary schools that accommodate more than 7000 children. PHC built and maintains more than 70 boreholes, providing access to drinking water in rural areas. The company also launched PHC Ventures, an initiative to support youth entrepreneurship in the areas of agriculture, health and renewable energy, mentoring young people to realize their business ambitions. PHC has established the PHC Foundation, a not-for-profit public interest institution dedicated to enhancing a positive and sustainable social and economic impact on communities established around its operational sites [63]. Clearly, mechanisms exist for PHC to comprehensively channel revenue from carbon credits toward the local communities that surround its plantations. The eligibility of such plantations can also be factored into national carbon reduction strategies and commitments.

8. Words of caution and potential applications

While this chapter argues in favor of breaking with the past and awarding carbon credits to oil palm plantations that sustainably manage their environmental and social responsibilities in a demonstrable manner [8, 9, 11], we also understand the concerns of those who seek to keep the Congo Basin, and other equatorial forests, in as natural state as possible. These forests are considered the "lungs of the Earth" (along with oceans) and massive carbon sinks [64], and any efforts to encourage those who disturb these systems threatens more idealist world views of tropical ecosystem function.

However, the peculiar situation of the DRC's component of the Congo Basin needs to be taken into consideration. The main access to the Congo Basin is along the River Congo and its tributaries, especially along its southern shore. Land disturbance penetrates about 50 km or so inland along these rivers, and beyond that are vast Guineo-Congolian humid forests rich in carbon and biodiversity [65]. Due to logistical constraints, palm oil mills need to be located near the Congo River or other major rivers that serve as supply corridors for goods needed for or products coming from industrial operations. Since, for operational efficiency, oil palm plantations need to be located near palm oil mills, it is recommended that sustainable palm oil development policies be limited to within 50 km of the major rivers of the Congo Basin, which is where most degraded forests are in the first place. To many who have never traveled to the Congo Basin, the world seems best served if these forests, their wildlife and inhabitants are protected from any further "disturbance," however, this non-disturbance may equate to many living in extreme poverty [1].

One may argue that funding oil palm plantations stands in direct contrast to investments that protect and restore nature and may even conflict with past biodiversity agreements [66]. Admittedly, given the limited scope of investment, carbon market funds directed to one sector (e.g., oil palm sequestration and decarbonizing) become less available to others (the protection and restoration of forests). In addition, any further investment in the Congo Basin is likely to improve transportation infrastructure that makes additional forests more accessible and exploitable, and this may be viewed as a leakage from measured carbon gains elsewhere. In this way, the greater appeal of natural forest carbon markets is understandable, and across the DR Congo, there are many forest concessions available for such protection. At the same time, forest carbon markets are subject to credibility issues, and this difficulty discourages investment. On the other hand, the carbon accrued within oil palm plantations is readily accessible and measurable, largely because it exists as a relatively homogeneous overstorey monocrop, based upon a large body of credible science identified in this chapter [27–38]. Verification of performance gains is simply more direct and less costly in oil palm plantations than in natural forests for a host of reasons.

What is lacking in support of oil palm vs. forest carbon in Central Africa are strong counterfactual baselines. Once protected, disturbed lands also accrue system C and reestablish plant and animal biodiversity at relatively high rates as they progress into mature secondary forests. There are too few pairwise comparisons of plantations vs. recovering forests, making it difficult to determine if carbon benefits are being ignored or overstated [18, 19]. On the other hand, less uncertainty exists among carbon market actors engagement with local communities concerning the equitable sharing of carbon benefits. Forest encroachment is by nature a furtive process, and it is difficult to monitor the changed behavior of forest stakeholders, whereas the communities within and surrounding remotely located oil palm plantations are much more organized and approachable, and the plantation itself is well positioned to deliver services that improve stakeholder lives and livelihood [1, 62].

In any event, better-establishing baselines and examining counterfactuals within the Congo Basin are important research priorities in the future. Along these lines, we hypothesize that "as the general public becomes better aware of the improved environmental and social corporate responsibility of Africa's palm oil producers in the future, carbon investors will consider them increasingly acceptable partners within mitigation strategies," and this will positively impact upon humanity's ability to mitigate climate change in a modest, but nonetheless important way into the near future.

9. Conclusions

This paper describes the carbon dynamics of oil palm plantations around the world, with special attention to explaining how oil palm plantations established in Africa for over a century, such as those of the Plantations et Huileries du Congo (PHC) in DR Congo, could qualify as a carbon offset provider. We believe this case is compelling and that much of the current condemnation of the sector is unjustified. The oil palm sector is admittedly controversial based upon a history of negative social and environmental impacts, but it also generates regular income for numerous large- and small-scale growers. The sector has grown more complex in terms of its institutional, social, ecological and environmental dynamics [67]. Recent efforts support the transition to a more sustainable oil palm value chain, with important contributions from both the public and private sectors, and several complementarities have emerged [68]. The greater involvement of civil society organizations strengthens this development.

Oil palm is highly productive and profitable, and worldwide oil demand is rising. At the same time, further oil palm development involves several social and environmental tradeoffs. The crop provides a way out of poverty but also makes communities vulnerable to exploitation [69]. It threatens biological diversity but at the same time potentially offers resources needed to protect adjacent forests. It offers renewable fuel but also threatens to increase global carbon emissions.

Optimizing these tradeoffs in a way that provides local, national and international benefits is the challenge before us. We believe that this challenge can be overcome and that the African palm oil, and the Congo basin in particular, could give rise to a well-regulated and appropriately documented palm oil industry in conformity with science-based criteria prioritizing reduction in carbon emissions and biodiversity conservation. Because of their positions deep within the humid forest zone of DR Congo, meteorological impacts associated with climate change, such as warming trends, changes in rainfall and extreme weather events, are less likely to threaten the viability of oil palm plantations in the Congo basin in the foreseeable future [1, 70]. There will be no substitution of this globally important crop as an adaptive option to climate change; rather, these plantations will continue to accrue biomass at a relatively high and manageable rate, making them potentially important carbon sinks. That many of these plantations were carved from humid forests that contained even more system carbon several decades ago should no longer be pertinent.

Rather, what is today relevant is that (1) this practice of forest conversion is now discouraged and discontinued, (2) management practices that allow for additional carbon storage within plantations are known and utilized and (3) future oil palm plantations are acquired and developed from less productive, disturbed lands rather than carbon-rich forests. The present risk is not the existence of plantations themselves but rather ignoring their potential for greater carbon storage as a mitigative response to the threat of climate change because of historical land management grievances. We cannot ignore the great contribution of oil palm plantations to meeting the social and economic needs of human settlements in the Congo Basin and across the African palm oil belt. Indeed, well-managed oil palm plantations can contribute to progress toward net zero emissions while at the same time uplifting millions of people out of poverty and providing good economic returns to stakeholders.

Acknowledgements

Useful insights into the difficulties in assigning carbon credits to palm oil producers were provided by Chris Okafor of the African Agricultural Leadership Institute. Dries Roobroeck of IITA-Kenya provided insights into biochar production. Plantations et Huileries du Congo (PHC) provided information that allowed for projections appearing in this paper to be generated. Angelique Kajibwami of the PHC Foundation provided information on corporate social responsibility. Discussions leading to the production of this paper were initiated through the DR Congo Agricultural Transformation Agenda, organized by the DRC Office of the President. The International Institute of Tropical Agriculture provided additional support for the development of this paper. The contributions of these individuals and organizations are gratefully acknowledged.

Author details

Paul L. Woomer[1]* and Mpoko Bokanga[2]

1 International Institute of Tropical Agriculture (IITA), Nairobi, Kenya

2 PHC Foundation, Kinshasa, DR Congo

*Address all correspondence to: plwoomer@gmail.com

References

[1] Bokanga M. Congo: The next frontier for the palm oil industry. In: Waisundara VY, editor. *Elaeis guineensis*-New Insights. London: IntechOpen; 2024. DOI: 10.5772/intechopen.114010

[2] Brandão F, Schoneveld G, Pacheco P, Vieira I, Piraux M, Mota D. The challenge of reconciling conservation and development in the tropics: Lessons from Brazil's oil palm governance model. World Development. 2021;**139**:105268

[3] Mosnier A, Boere E, Reumann A, Yowargana P, Pirker J, Havlík P, et al. Palm oil and likely futures: Assessing the potential impacts of zero deforestation commitments and a moratorium on large-scale oil palm plantations in Indonesia. CIFOR; May 2017. No. 177. DOI: 10.17528/cifor/006468 [Accessed: August 2024]

[4] Austin KG, Mosnier A, Pirker J, McCallum I, Fritz S, Kasibhatla PS. Shifting patterns of oil palm driven deforestation in Indonesia and implications for zero-deforestation commitments. Land Use Policy. 2017;**69**:41-48

[5] Sipayung T. Palm oil industry will become a net carbon sink. Journal of Palm Oil Environment. 2021;**II**(12/47/2021):575-579. Available from: https://palmoilina.asia/jurnalkelapa-sawit/net-zero-emissions/#9-volii-no-47122021-palm-oil-industry-willbecome-a-net-carbon-sink

[6] Tropical Forest Alliance. TFA African Palm Oil Initiative was held in Brazzaville; 8-11 Oct 2018. Available from: https://www.tropicalforestalliance.org/en/news-and-events/news/tfa-2020-africapalm-oil-initiative-5th-regional-meeting [Accessed: August 2024]

[7] Mobateli A. Congo-Kinshasa: Le Pays va Relancer Plus de 2.000 Plantations Zairianisées et Abandonnées. Nairobi: Le Potentiel, AllAfrica Global Media; 2013. Available from: https://fr.allafrica.com/stories/201302090400.html

[8] RSPO. Principles and Criteria for the Production of Sustainable Palm Oil. Kuala Lumpur: Roundtable on Sustainable Palm Oil; 2018. Available from: https://rspo.org/resources/certification/rspo-principles-criteria-certification

[9] RSPO. Impact Update 2023. Kuala Lumpur, Malaysia: Roundtable on Sustainable Palm Oil; 2023. 34 p. Available from: https://rspo.org/wp-content/uploads/Impact-Update-2023.pdf

[10] Popkin M, Reiss-Woolever VJ, Turner EC, Luke SH. A systematic map of within plantation oil palm management practices reveals a rapidly growing but patchy evidence base. PLOS Sustainability and Transformation. 2022;**1**(7):e0000023. DOI: 10.1371/journal.pstr.0000023

[11] Weng CK. Best-developed practices and sustainable development of the palm oil industry. Journal of Oil Palm Research. 2005;**17**:124-135. Available from: http://jopr.mpob.gov.my/bestdeveloped-practices-and-sustainabledevelopment-of-the-oil-palm-industry/

[12] Munar DA, Ramírez-Contreras N, Rivera-Méndez Y, Garcia-Nuñez JA, Romero HM. Carbon footprint management for a sustainable oil palm crop. In: Ren J editor. Advances of Footprint Family for Sustainable Energy and Industrial Systems. Switzerland: Springer Nature; 2022. pp. 93-110

[13] Butler RA, Koh LP, Ghazoul J. REDD in the red: Palm oil could undermine carbon payment schemes. Conservation Letters. 2009;**2**(2):67-73

[14] Germer J, Sauerborn J. Estimation of the impact of oil palm plantation establishment on greenhouse gas balance. Environment, Development and Sustainability. 2008;**10**:697-716

[15] Hashim Z, Subramaniam V, Harun MH, Kamarudin N. Carbon footprint of oil palm planted on peat in Malaysia. The International Journal of Life Cycle Assessment. 2018;**23**:1201-1217

[16] Carlson KM, Curran LM, Asner GP, Pittman AM, Trigg SN, Adeney JM. Carbon emissions from forest conversion by Kalimantan oil palm plantations. Nature Climate Change. 2013;**3**(3):283-287

[17] Burton ME, Poulsen JR, Lee ME, Medjibe VP, Stewart CG, Venkataraman A, et al. Reducing carbon emissions from forest conversion for oil palm agriculture in Gabon. Conservation Letters. 2017;**10**(3):297-307

[18] Dislich C, Keyel AC, Salecker J, Kisel Y, Meyer KM, Auliya M, et al. A review of the ecosystem functions in oil palm plantations, using forests as a reference system. Biological Reviews. 2017;**92**(3):1539-1569

[19] Fitzherbert EB, Struebig MJ, Morel A, Danielsen F, Brühl CA, Donald PF, et al. How will oil palm expansion affect biodiversity? Trends in Ecology & Evolution. 2008;**23**(10):538-545

[20] Wilcove DS, Koh LP. Addressing the threats to biodiversity from oil-palm agriculture. Biodiversity and Conservation. 2010;**19**:999-1007

[21] KotowskaMM,LeuschnerC,TriadiatiT, Meriem S, Hertel D. Quantifying above- and belowground biomass carbon loss with forest conversion in tropical lowlands of Sumatra (Indonesia). Global Change Biology. 2015;**21**(10):3620-3634

[22] Ramos HMN, Vasconcelos SS, Kato OR, et al. Above- and belowground carbon stocks of two organic, agroforestry-based oil palm production systems in eastern Amazonia. Agroforestry Systems. 2018;**92**:221-237

[23] Sanquetta CR, Sylvio PÃ, Dalla Corte AP, Behlin A, Sanquetta MNI. Quantifying biomass and carbon stocks in oil palm (Elaeis guineensis Jacq.) in Northeastern Brazil. African Journal of Agricultural Research. 2015;**10**(43):4067-4075

[24] Singh SL, Sahoo UK, Kenye A, Gogoi A. Assessment of growth, carbon stock and sequestration potential of oil palm plantations in Mizoram, Northeast India. Journal of Environmental Protection. 2018;**9**(9):912

[25] PASPI-Monitor. Palm oil industry will become a net carbon sink. Palm Oil Journal. 2021;**II**(47):575-579. Available from: https://palmoilina.asia/wp-content/uploads/2021/12/2.47.-PALM-OIL-INDUSTRY-WILL-BECOME-A-NET-CARBON-SINK-2.pdf

[26] Lamade E, Bouillet JP. Carbon storage and global change: The role of oil palm. OCL. 2005;**12**(2):155-160

[27] Lewis K, Rumpang E, Kho LK, McCalmont J, Teh YA, Gallego-Sala A, et al. An assessment of oil palm plantation aboveground biomass stocks on tropical peat using destructive and non-destructive methods. Scientific Reports. 2020;**10**(1):2230

[28] Pulhin FB, Lasco RD, Urquiola JP. Carbon sequestration potential of oil

palm in Bohol, Philippines. Ecosystems and Development Journal. 2014;**4**(2):14-19

[29] Aranda-Arguello R, Ley-de Coss A, Arce-Espino C, Pinto-Ruiz R, Guevara-Hernández F, Raj-Aryal D. Carbon sequestration in aerial biomass of the oil palm in Chiapas, Mexico. Agronomía Mesoamericana. 2018;**29**(3):629-637

[30] Henson IE, Ruiz R, Romero HM. The greenhouse gas balance of the oil palm industry in Colombia: A preliminary analysis. I. Carbon sequestration and carbon offsets. Agronomía Colombiana. 2012;**30**(3):359-369

[31] Henson IE, Ruiz R, Romero HM. The greenhouse gas balance of the oil palm industry in Colombia: A preliminary analysis. II. Greenhouse gas emissions and the carbon budget. Agronomía Colombiana. 2012;**30**(3):370-378

[32] Henson IE, Betitis T, Tomda Y, Chase LD. The estimation of frond base biomass (FBB) of oil palm. Journal of Oil Palm Research. 2012;**24**:1473-1479

[33] Frazão LA, Paustian K, Cerri CEP, Cerri CC. Soil carbon stocks under oil palm plantations in Bahia State, Brazil. Biomass and Bioenergy. 2014;**62**:1-7

[34] Ahn BJ, Han GS, Choi DH, Cho ST, Lee SM. Assessment of the biomass potential recovered from oil palm plantation and crude palm oil production in Indonesia. Journal of the Korean Wood Science and Technology. 2014;**42**(3):231-243

[35] Migolet P, Goïta K, Ngomanda A, Mekui Biyogo AP. Estimation of aboveground oil palm biomass in a mature plantation in the Congo Basin. Forests. 2020;**11**(5):544

[36] Alometri PTM, Korom A, Phua M, Matsuura T. Relationships between crown size and aboveground biomass of oil palms: An evaluation of allometric models. Sains Malaysiana. 2016;**45**(4):523-533

[37] Ramos-Escalante G, Ley de-Coss A, Arce-Espino C, Escobar-España JC, Raj-Aryal D, Pinto-Ruiz R, et al. Alometric equations for estimating biomass and carbon in oil palm (*Elaeis guineensis* Jacq.) in the humid tropic of Chiapas, Mexico. Agrociencia. 2018;**52**(5):671-683

[38] Aholoukpè H, Dubos B, Flori A, Deleporte P, Amadji G, Chotte JL, et al. Estimating aboveground biomass of oil palm: Allometric equations for estimating frond biomass. Forest Ecology and Management. 2013;**292**:122-129

[39] Pulunggono HB, Anwar S, Mulyanto B, Sabiham S. Decomposition of oil palm frond and leaflet residues. AGRIVITA Journal of Agricultural Science. 2019;**41**(3):524-536

[40] Rahman N, Giller KE, de Neergaard A, Magid J, van de Ven G, Bruun TB. The effects of management practices on soil organic carbon stocks of oil palm plantations in Sumatra, Indonesia. Journal of Environmental Management. 2021;**278**:111446

[41] Rivera-Méndez YD, Rodríguez DT, Romero HM. Carbon footprint of the production of oil palm (*Elaeis guineensis*) fresh fruit bunches in Colombia. Journal of Cleaner Production. 2017;**149**:743-750

[42] Hong WO. Review on carbon footprint of the palm oil industry: Insights into recent developments. International Journal of Sustainable Development and Planning. 2023;**18**:447-455. DOI: 10.18280/ijsdp.180213

[43] Patthanaissaranukool W, Polprasert C. Carbon mobilization in oil palm plantation and milling based on a carbon-balanced model-a case study in Thailand. EnvironmentAsia. 2011;**4**(2):17-26

[44] Patthanaissaranukool W, Polprasert C, Englande AJ Jr. Potential reduction of carbon emissions from Crude Palm Oil production based on energy and carbon balances. Applied Energy. 2013;**102**:710-717

[45] Moreno García JE, Martínez LA, Belalcázar Cerón LC, Rojas NY. Corporate carbon footprint of a palm oil mill. Tecciencia. 2018;**13**(24):1-10

[46] Jamaludin NF, Ab Muis Z, Hashim H. An integrated carbon footprint accounting and sustainability index for palm oil mills. Journal of Cleaner Production. 2019;**225**:496-509

[47] Zakaria MR, Farid MAA, Andou Y, Ramli I, Hassan MA. Production of biochar and activated carbon from oil palm biomass: Current status, prospects, and challenges. Industrial Crops and Products. 2023;**199**:116767

[48] Bindar Y, Steven S, Kresno SW, Hernowo P, Restiawaty E, Purwadi R, et al. Large-scale pyrolysis of oil palm frond using two-box chamber pyrolyzer for cleaner biochar production. Biomass Conversion and Biorefinery. 2024;**14**(5):6421-6434

[49] Abnisa F, Arami-Niya A, Daud WW, Sahu JN, Noor IM. Utilization of oil palm tree residues to produce bio-oil and bio-char via pyrolysis. Energy Conversion and Management. 2013;**76**:1073-1082

[50] Umar HA, Sulaiman SA, Meor Said MA, Gungor A, Shahbaz M, Inayat M, et al. Assessing the implementation levels of oil palm waste conversion methods in Malaysia and the challenges of commercialisation: Towards sustainable energy production. Biomass and Bioenergy. 2021;**151**:106179. DOI: 10.1016/j.biombioe.2021.106179

[51] Harson SS, Grundman P, Lau LH, Hansen A, Salleh MAM, Meyer-Aurich A, et al. Energy balances, greenhouse gas emissions and economics of biochar production from palm oil empty fruit bunches. Resources, Conservation and Recycling. 2013;**77**:108-115

[52] Liew RK, Nam WL, Chong MY, Phang XY, Su MH, Yek PNY, et al. Oil palm waste: An abundant and promising feedstock for microwave pyrolysis conversion into good quality biochar with potential multi-applications. Process Safety and Environmental Protection. 2018;**115**:57-69

[53] Razali N, Kamarulzaman NZ. Chemical characterizations of biochar from palm oil trunk for palm oil mill effluent (POME) treatment. Materials Today Proceedings. 2020;**31**:191-197

[54] Kong SH, Loh SK, Bachmann RT, Rahim SA, Salimon J. Biochar from oil palm biomass: A review of its potential and challenges. Renewable and Sustainable Energy Reviews. 2014;**39**:729-739

[55] Hwong CN, Sim SF, Kho LK, The YA, Harrold LD, Chua KH, et al. Effects of biochar from oil palm biomass on soil properties and growth performance of oil palm seedlings. Journal of Sustainability Science and Management. 2022;**17**(4):183-200. DOI: 10.46754/jssm.2022.4.014

[56] Kätterer T, Roobroeck D, Andrén O, Kimutai G, Karltun E, Kirchmann H, et al. Biochar addition persistently increased soil fertility and yields in maize-soybean rotations over 10 years in sub-humid regions of Kenya. Field Crops Research. 2019;**235**:18-26. DOI: 10.1016/j.fcr.2019.02.015

[57] Wang J, Xiong Z, Kuzyakov Y. Biochar stability in soil: Meta-analysis of decomposition and priming effects. GCB Bioenergy. 2016;**8**:512-523. DOI: 10.1111/gcbb.12266

[58] Maximilian P. L'utilisation du Biochar en Agroforesterie Recommandé pour une Agriculture Soutenue et Propre. 2015. Available from: https://www.mediacongo.net/article-actualite-13936_l_utilisation_du_biochar_en_agroforesterie_recommande_pour_une_agriculture_soutenue_et_propre.html

[59] Nyami ABL, Sudi CK, Lejoly J, Kalala PM. Effet du biochar sur la rétention et la disponibilité en eau et éléments minéraux pour les plantes dans un sol sableux de Kinshasa. Revue Africaine d'Environnement et d'Agriculture. 2022;**5**(3):19-29. Available from: http://www.rafea-congo.com/pages/lecture1.php?id_article=183

[60] de Wet L. Baseline Vegetation and Flora Assessment, Lokutu Concession, Feronia, DRC. South Africa: LD Biodiversity Consulting, Digby Wells; 2015. 61 pp

[61] Furuichi BT, Idani G, Kimura D, Ihobe H, Hashimoto C, editors. Bonobos and the People of Wamba: 50 Years of Research. Singapore: Springer Nature; 2023. 589 pp

[62] PHC. Sustainability: Social Impact and Community Development Program. Kinshasa: Plantations et Huileries du Congo (PHC); 2024. Available from: https://phc-congo.com/en/impacts-sociaux/

[63] PHC Foundation. Available from: www.fondation-phc.org

[64] Mackey B, Kormos CF, Keith H, Moomaw WR, Houghton RA, Mittermeier RA, et al. Understanding the importance of primary tropical forest protection as a mitigation strategy. Mitigation and Adaptation Strategies for Global Change. 2020;**25**(5):763-787

[65] White F. The vegetation of Africa Natural Resources Research. Vol. 20. Paris: UNESCO; 1983. 356 pp

[66] Burgass MJ, Larrosa C, Tittensor DP, Arlidge WN, Caceres H, Camaclang A, et al. Three key considerations for biodiversity conservation in multilateral agreements. Conservation Letters. 2021;**14**(2):e12764

[67] Pacheco P, Gnych S, Dermawan A, Komarudin H, Okarda B. The Palm Oil Global Value Chain: Implications for Economic Growth and Social and Environmental Sustainability. Bogor, Indonesia: CIFOR; 2017

[68] Pacheco P, Schoneveld G, Dermawan A, Komarudin H, Djama M. Governing sustainable palm oil supply: Disconnects, complementarities, and antagonisms between state regulations and private standards. Regulation & Governance. 2020;**14**(3):568-598

[69] Sheil D, Casson A, Meijaard E, van Noordwijk M, Gaskell J, Sunderland-Groves J, et al. The impacts and opportunities of oil palm in Southeast Asia: What do we know and what do we need to know? In: Occasional Paper no. 51. Bogor, Indonesia: CIFOR; 2009

[70] Peterson RRM. Longitudinal trends of future climate change and oil palm growth: Empirical evidence for tropical Africa. Environmental Science and Pollution Research. 2021;**28**:21193-21203

Section 2

Adaptation and the Reduction of the Consequences of Climate Impacts

Chapter 3

Impact Prospect of Heatwaves in the Midst of Climate Instability in Europe

Julian Schlubach

Abstract

Heatwaves have a growing impact on humans, ecosystems, and agriculture across Europe, while soil moisture and land cover represent key mitigation mechanisms endangered by the ongoing climatic change. Handling the situation as it evolves, with strong constraints on natural resources, is expected to become a major challenge, while health, ecosystems, and production systems will be under increased pressure. This chapter aims to present a state of research regarding the interaction between land cover and local climate in the context of global warming. The work is based on previous research and reviews completed for the present chapter. This opens further research perspectives assessing the soil-air-water interactions and climate, which is critical considering territorial planning. Heatwaves increased in frequency and intensity at the turn of the century. Global warming mechanisms affecting local warming and heat repartition can be somewhat influenced. From this perspective, permanent land cover, also endangered by climate change, plays a crucial role in mitigating climate at the local level. A detailed assessment of the change occurring in the Mediterranean region will be conducive to feeding further thoughts regarding upcoming challenges across Europe.

Keywords: heatwave, climate change, global warming, soil moisture, heat tolerance, land cover, forest, resilience, mitigation, temperature thresholds, high temperatures, evapotranspiration

1. Introduction

Notably, 2023 has been the hottest year since meteorological data were recorded, and 2024 [1] is about to beat the record. Since 2020, 3 years have been the hottest on record [2]. Heatwaves over Europe are becoming increasingly frequent and gaining intensity.

The Green House effect is a mechanism that allows Earth to keep its surface temperature within a range compatible with the development of life. The warming happens mainly because the atmosphere absorbs infrared wavelengths reemitted by the ground. There is nothing new in the fact that some gas in the atmosphere composition plays a prominent role in this regard. The composition of the atmosphere has itself changed over Earth's geological times. Fluctuations, including cold and hot episodes, have occurred over Earth's history. Past fluctuations can be assessed through geological analysis and, for the quaternary era, through ice probes analysing the gas composition and isotope proportions. The worst event occurred about 252 million years ago at the end of the Permian era [3, 4], leading to Earth's worst life mass extinction. The massive eruption of an area as big as Alaska, now located in Siberia, led to the massive release of greenhouse gas (GHG) and a brutal temperature change. Considering less brutal fluctuations which prevailed throughout Earth's history, the main difference between ongoing global warming and past events is related to the amplitude and speed of change.

Different factors intervene in Earth's surface temperature fluctuation. The Sun emits more energy now than it did one billion years ago, and it also has cycles of activity that affect Earth [5]. The nature of the soil also modifies the radiative balance on Earth; oceans have a cooling effect through evaporation, as well as plants through evapotranspiration, while ice contributes to a direct reflection to Space. Besides, air, water, and soil interactions intervene in the repartition of energy balance at the globe's surface. Oceanic currents play an important role in temperature repartition. An insight into meteorological data over the past provides a useful perspective regarding the ongoing challenge. Even though the thermodynamic regulation of the Earth's atmosphere is a dynamic equilibrium that may evolve to some extent under global warming, both phenomena require clear distinction.

The World Meteorological Organisation defines heatwaves as 'a period where local excess heat accumulates over a sequence of unusually hot days and nights' [6]. Thus, thresholds regarding the maximum temperature during the day and minimal temperature at night are considered. In most cases, a heatwave will be defined as an abnormal deviation from average temperatures observed over a thirty-year reference period. Considering plants' physiological limits like wheat, events involving a few days in a row with day temperatures over 30°C and/or tropical nights with temperatures not dropping below 20°C can be considered a definition. The definition can be much less restrictive, like the criteria established in 1947 by the French Meteorological Organisation [7], considering that temperature should be 1 day above 25.3°C and greater than or equal to 23.4°C for at least 3 days. Those episodes are not new, but an increase in the frequency or intensity of heatwave events is expected in the background of global warming. The role of land cover in climate attenuation at the local level and the local interaction with global warming is being acknowledged [8], as is the need for additional research to feed territorial management decisions. While new thresholds are bypassed, a global perspective on the impact on human health, forests, energy, agriculture, and retroactions between land cover and climate requires more attention.

The chapter is divided into five main parts (two to six). The methodology applied to the review is described at first. The following sections present a historical perspective illustrating the change in Ref. temperatures accompanying the heatwave concept, a description of the mechanisms triggering heatwaves, distinguishing earth's global energetical balance from heat fluxes repartition mechanisms, the observed impacts and dangers faced by the vegetation land cover, and the projections from models (**Figure 1**).

The conclusion aims to highlight key issues and define research to be conducted at the local level to clarify further the understanding of the challenges faced and what can be done.

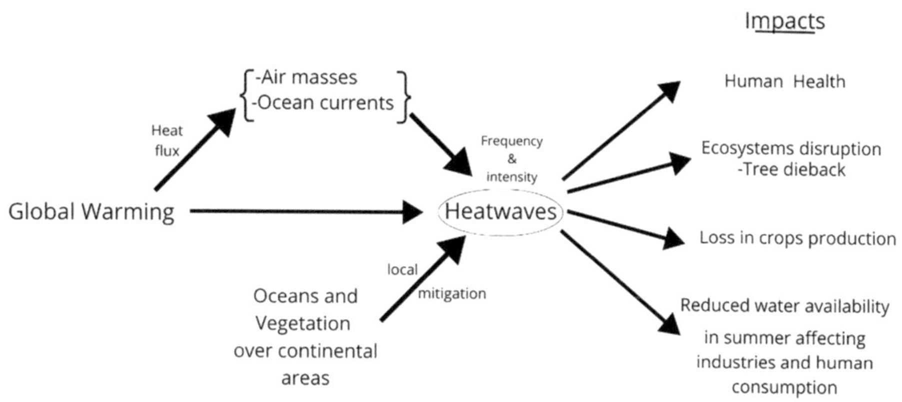

Figure 1.
Heatwaves influencing factors and impact—© Julian Schlubach, 2024.

2. Methodology

The present chapter continues previous work on the impact of global warming on field crops, considering plants' physiological limits [9] and regarding the role of ecosystem services in adaptation to global warming [10]. It is opening perspectives on complementary research regarding territorial planning and mitigation measures. The aim is to provide a preliminary assessment of the tendency regarding heatwaves, their impact, and the extent to which the ecosystem services we may rely on from a climate mitigation perspective might be endangered. The proposed approach implies cross-cutting perspectives involving different research units and institutions. The present chapter also intends to establish a preliminary basis for discussions addressing the global warming challenge at the territorial level.

The core elements addressed under sections 4 to 6 are based on academic peer-reviewed research partly collected in the framework of previous research from the author [9, 10] and partly on additional references collected through a screening of research publication platforms. The climate and impact parts are based on peer-reviewed publications retrieved through research platforms like Science Direct, Elsevier, Springer, ResearchGate, or Nature. Publications accessible free of charge have been privileged, and some others have been purchased. The climate tendency described in Section 3 is mainly based on grey literature; Google and Wikipedia have been used for this purpose, and each reference has been cross-checked. Section 3 thus illustrates how the perception of temperatures and heatwaves has evolved while changes occur. The overall references used are not exhaustive but intend to illustrate important points that require attention, according to the state of the art. The analysed climate data related to meteorological references, including projections, in different European countries have been extracted from the World Bank climate open portal [11]. Meaningful available data to illustrate observed trends have been selected and extracted, even though averages are not optimal for analysis. The retrieved average temperatures have been put into perspective, and explanations regarding the interpretation are provided. The aim is to illustrate the main trends and challenges.

3. Historical perspective

Following the last glaciation period, the Holocene warming period reached a maximum estimated to have occurred 6500 years ago [12]. Different models show a global temperature decrease of 0.5°C until the industrial era. Thus, over time, Earth has slowly evolved towards a cooler climate. However, this trend stopped in the nineteenth century, with the industrial era. Although meteorological data do not allow much insight into recorded temperature beyond 1850, historical descriptions provide a perspective on the occurrence of extreme events. Some historically documented heatwaves were recorded across Europe, such as in 1540 [13], with extreme drought and hot temperatures over the year, and in 1808 [14] and 1858 [15] in the United Kingdom. Dendrology, assessing tree rings, provides a complementary insight into local climatic events, which have resulted in hydric stress in the past [16].

Over the twentieth century, at least nine heatwaves were reported in Europe, particularly in 1911, 1947, 1952, 1976, 1983, 1987, 1990, 1994, and 1997. In 1911, a heatwave across France and the United Kingdom (UK) beat a record of 36°C at the beginning of August for the UK [17]. In 1947, between the 27th of July and the 5th of August, maximal temperatures exceeded 40°C and a record of 37.6°C in Paris [18]. In 1952, Romania experienced a long hot day episode in August [19]; temperatures peaked at 40°C on the 17th, while abnormally high temperatures were also recorded at the beginning of October. The 1976 heatwave over North-Western Europe [20] lasted two consecutive weeks; day temperatures in the UK did not drop below 32°C, peaking at 35.9°C on July 3rd. In 1983 [21], a heatwave affected North-Western Europe over a few weeks in July, with temperatures reaching 38°C. In 1987, a one-week-long hot event affected an area extending from Italy in the South to the United Kingdom and Germany in the North. Registered day temperatures reached 40°C and barely dropped below 30°C at night [22]. Over a few days at the beginning of August 1990, temperatures peaked at 37°C in the UK [23] and reached 40°C in France [24]. A heatwave affected especially Poland between July and August 1994 [25], with temperatures reaching 39.5°C. In 1997, the United Kingdom faced one of the hottest summers on record, and the period from May 1997 to September 1998 was recorded as the hottest period worldwide on record [26].

Recorded heatwaves became more frequent at the beginning of the twenty-first century. At least eleven occurrences were recorded for Europe over twenty-three years: 2003, 2006, 2007, 2010, 2011, 2013, 2015, 2017, 2018, 2022, and 2023. The frequency is more than four times higher than in the past century. Besides, the intensity of the recorded events is unprecedented (**Figure 2**).

In the United Kingdom, the record of 36.7°C set in 1911 was only beaten in 1990, when temperatures reached 37°C; hotter records have succeeded each other since 2000, reaching a record of 40.3°C on 19 July 2022 [27]. The same acceleration in the frequency and intensity of hot days is observed in France. The study of average temperatures from thirty meteorological stations since 1947 shows a steady increase in temperature records, with a considerable acceleration since the nineties [28]. Heatwaves in 2003 [29] and 2006 [30] affected most of Western Europe, breaking new temperature records with 47°C in the South of Portugal in 2003 and 40°C in Paris in 2006. The 2007 heatwave affected mainly the South of Europe, with heat extremes reaching 45°C in Bulgaria [31]. In June 2010 [32], a heatwave from the Sahara hit the East of Europe. In July 2015 [33], the West, including England, was affected, with temperatures reaching 37°C. In June 2017 [34], a heatwave affected Greece and the

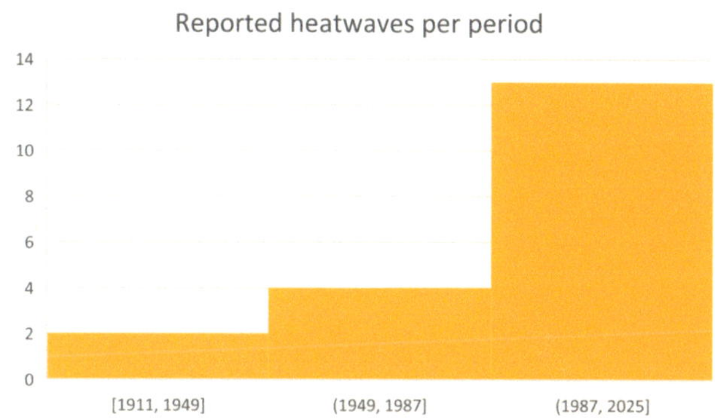

Figure 2.
Number of heatwaves reported in media between 2011 and 2023 per 38-year period—© Julian Schlubach, 2024.

Balkans. In the summer of 2018 [35], Northern and Central Europe faced extreme heat and drought, and in July 2022 [36], most of Western Europe faced a heatwave. The United Kingdom experienced abnormal temperatures in October 2011 [37], July 2013 [38], and June 2017 [39]. In May 2022, a heatwave mainly affected Spain and France, among other European countries. The reached temperatures were the highest recorded in France since meteorological records exist, and a height of 41°C was recorded in Sevilla in Spain [36]. Summer temperatures in 2022 ranged from 2 to 4 degrees above the average registered over the preceding thirty-year reference period. In July 2023 [40], a heatwave mostly affected the Mediterranean region. In 2023, Temperatures were above average for 11 months of the year.

While heatwaves are not a new phenomenon in Europe, hot days lasting a few days in a row are becoming more frequent, with increasingly high recorded temperatures. Although it could be argued that this may enter a longer-term cycle, the occurrence of those events has to be placed in the context of the global climatic trend [26].

4. Mechanisms influencing heatwaves occurrence and intensity

Different mechanisms of energy exchanges between the atmosphere, oceans, and soils influence seasonal heat repartition on Earth. Air mass movement can induce rapid changes or increase the intensity of cold events or heatwaves. Global warming increases the overall energy intake at the Earth's surface, modifying the local albedo and energy balance and possibly modifying some pre-existing dynamic equilibriums. At the ground level, the land surface radiative balance and soil moisture influence the climate. Thus, global warming increases the probability of heatwaves, while energy exchange dynamic mechanisms can intensify such events.

4.1 Global warming

At equilibrium, the earth reemits to space the same quantity of energy it receives. Until the nineteenth century, the climate was cooling [11], which means that the quantity of energy reemitted was slightly more important than the energy received. With the industrial era, the trend has been progressively reversed. Between 2000 and 2015, NASA reported an average net influx of 0.7 Watts per square metre (**Figure 3**) [41].

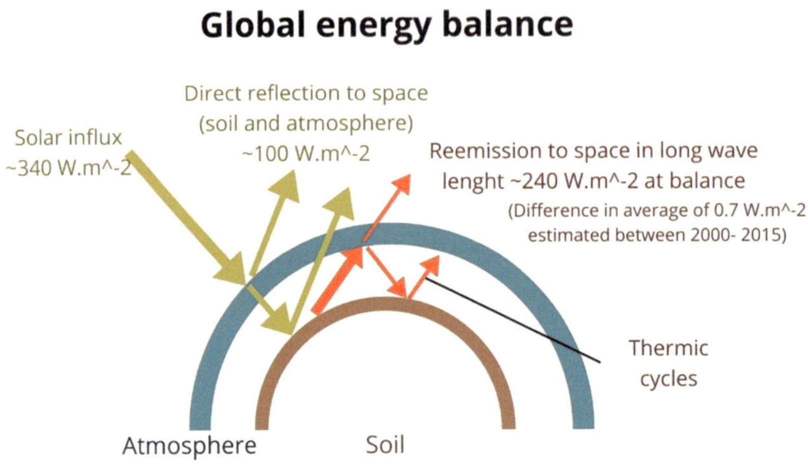

Figure 3.
Earth energy balance equilibrium (data from NASA [40])—© Julian Schlubach, 2024.

The difference in climate between Upper Normandy on the Atlantic North Coast of France and Provence on the Mediterranean Coast in the South of France is mainly due to the distribution of precipitation and the amplitude of temperature variations during the year. The average yearly precipitation over the 1991–2020 period was 823 mm for the Basse-Normandie Region and 849 mm in the Provence-Alpes-Côte d'Azur (PACA) region, with an average minimal temperature of 7.71°C in Basse-Normandie and 7.57°C in PACA, while average high temperatures are respectively of 15.19 and of 15.14°C [11]. However, these values also reflect that for average data, a minimal difference in measurement can represent a significant difference in seasonal and daily climatic variations for a given territory. Thus, the evolution of average annual temperatures for France, which oscillated around 13–14°C during the first half of the twentieth century to reach 16–17°C during the 2010–2020 decade, represents a development with observed impacts fuelling concerns.

In 2022, the average yearly mean temperature was 3.84°C for Sweden, 9.7°C for Poland, 10.11°C for the United Kingdom, 10.83°C for Germany [11], 13.16°C for France, 14.38°C for Italy, 14.88°C for Greece, and 15.3°C for Spain. The average value does not reflect the day and night and seasonal temperature variations. It is thus an indicator to handle with caution when considering vulnerability to climate change. In higher latitudes, heatwaves and droughts in summer are increasingly endangering ecosystems, even though temperatures are dropping to much lower levels than in Southern countries over the winter. Nevertheless, in the short term, the increase in average temperature does not have the same significance and impact in countries facing already higher temperatures. Besides, models present a slightly stronger summer temperature increase in Southern countries. However, considering the projected considerable increase in average temperature, tolerance thresholds in temperatures of ecosystems are likely to be increasingly frequently bypassed in all parts of Europe over the summer period. Continental areas face a higher challenge in this regard. Heatwaves are also likely to happen more often in spring and fall, with intra-seasonal, high-temperature variations, which could destabilise field crops.

4.2 Heat repartition: Intensifying and regulation mechanisms

Beyond Global Warming, three main factors related to energy repartition are presented hereafter: the jet stream under the polar vortex influence, the soil moisture, and the Atlantic Meridional Overturning Circulation (AMOC). In the present case, the cloud albedo variation is not addressed as a main factor of variation, as heatwaves are also correlated with strong anticyclones. The role of the AMOC and the way it might evolve is a mechanism regarding energy repartition on the earth's surface to be considered in a longer-term perspective.

4.3 The polar vortex and the double jets

The polar jet stream is located at 50–60 degrees northern latitude. The subtropical jet stream is located around 30 degrees North at the level of North Africa. The jet streams are located, at about ten kilometres altitude, at the limit of the troposphere and of the Stratosphere. Over the polar circle, the polar vortex circles up in the stratosphere, influencing the jet stream below. Those high-altitude streams influence air masses on the ground, triggering North-South, as well as low-high-altitude air masses exchanges, which can result in rapid changes and extreme meteorological events on the ground. The polar vortex is occasionally split into two vortexes circling in the form of an eight, disrupting the jet stream below, South and Northward (**Figure 4**).

Abnormal hot events have been linked to the polar vortex disruption with a double jet stream [42]. A strong correlation has been established between those conditions and a stable, strong anticyclone over the North Atlantic, involving heat events over the United Kingdom, favouring hot and dry weather over Western Europe. The record of observations of the polar vortex disruption and double jet phenomenon covers an insufficient period to conclude an increase in the frequency of occurrence beyond possible natural cycles. However, models show that while over the Arctic Ocean, heat is absorbed by melting ice and water warming, the surrounding continental area's rapid warming contributes to the shift of the equator to pole gradient towards the North Arctic Circle [42]. This ongoing change could favour double jet events. Warmer

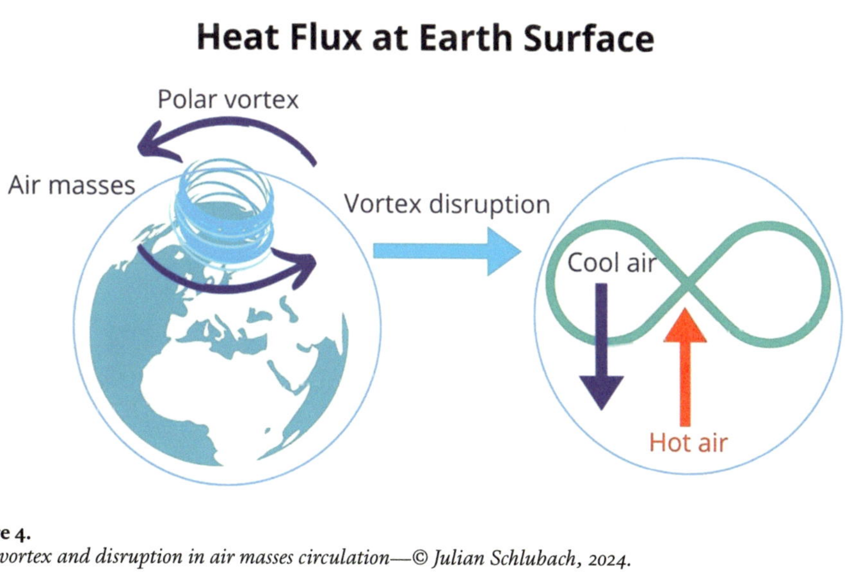

Figure 4.
Polar vortex and disruption in air masses circulation—© Julian Schlubach, 2024.

land areas in the Arctic could also favour the persistence of heatwave events. Double jet persistence correlates 30% with the variability in European heatwaves [43]. Thus, it is an important factor, but not the sole one.

4.3.1 Soil moisture

Soil moisture linked to soil temperature feedback influence on the local climate also plays an important role [43]. Global warming adds to the intensity (height of temperatures) and the cumulative intensity (cumulative excess temperature over a consecutive period of high-temperature event) (**Figure 5**).

Early Summer soil moisture strongly correlates with heatwaves observed over Western Europe from 1980 to 2011 [43]. This is also the case around the Mediterranean Basin [44]. Water evaporation and evapotranspiration absorb energy, resulting in a cooling effect. While the soil surface dries up quicker, vegetation can mobilise water volumes stored in lower earth layers. The maximum volume of water stored depends on the nature of the soil. The effective volume of stored water is measured by an index of the ratio of the Volume of Water in a volume of soil [43]. The soil can release water until a wilting point beyond which the earth is too dry. The available water volume for evapotranspiration is the difference between the volume of stored water and the wilting point. If the early summer period is dry, the potential for evapotranspiration over summer is reduced. In the Mediterranean region, soil moisture limits evapotranspiration in the summer, while radiation limitation is the main limiting factor in the Arctic tundra [45]. Along the coats, the land benefits from the cooling effect of the evaporation over water masses. The climate rapidly changes when entering inland, and soil moisture becomes a determining factor. Forest cover plays a crucial role in this regard, allowing plants to mobilise bigger volumes of water from the soil [46]. The change occurs over small distances. In the Normandie region of France, three climatic zones are distinguished from the coastline while moving inland. The 'oceanic climate benefits from cooler temperatures and more rain, while the 'altered oceanic climate' and the 'degraded oceanic climate' more inland present drier and hotter features [47]. In Central and Eastern Europe, the scope for climate

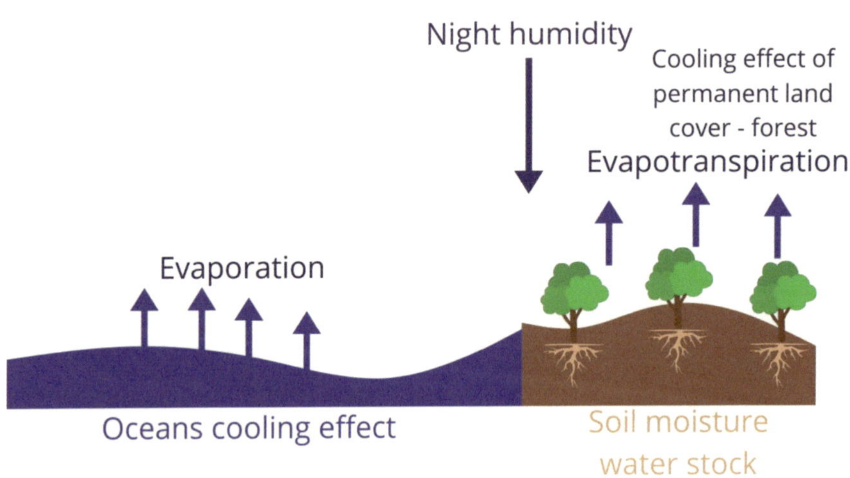

Figure 5.
Evaporation and evapotranspiration cooling effect at the local level—© Julian Schlubach, 2024.

regulation under global warming conditions is even more limited as soil moisture potential fades. A permanent land cover is of strategic importance for those areas.

4.3.2 Atlantic meridional overturning circulation (AMOC)

The European climate is influenced by the Atlantic Meridional Overturning Circulation (AMOC) [48], which transfers hot water from the Gulf of Mexico, warming the European coasts and atmospheric circulation, which turns according to the earth's rotation. The AMOC is linked to the Coriolis forces resulting from earth rotation and may thus not disappear. However, changes in convection between deep ocean and superficial water may influence surface temperatures. A less strong current could also affect the heat transfer from the Gulf of Mexico towards the European coasts. The accelerated melting of the polar ice cap is modifying the water salinity, which may affect the deep-water flux along the North American coast. A current slowdown would affect heat repartition on both sides of the Atlantic and ocean-atmosphere dynamics. However, this may only marginally affect the tendency towards hotter summer and heatwaves occurrence. From this perspective, it may be noted that the cold water flowing down from the Arctic along the Canadian coast does not stop the occurrence of heatwaves, droughts, and subsequent devastating wildfires over the Canadian North.

5. Impact of heatwaves

Different parameters need to be taken into account when considering the impact of the exposure to high temperatures. The impact of heatwaves on ecosystems depends on the temperature intensity and prevailing drought conditions. Humidity, when associated with high temperatures and, in some cases, pollution, has a deadly effect on living beings. Heatwaves over the past two decades have resulted in a considerable increase in deaths, especially among elderly persons. Besides, changes in temperature and humidity are increasingly destabilising ecosystems. While heatwaves are succeeding, their impact can be observed, even though looking into the past will not properly inform either about upcoming changes.

5.1 Impact on human health and death toll

The 1911 heatwave has not been exceptional compared to the 1990–2020 reference periods. However, summer temperatures have been far above the recorded temperatures between 1870 (the first meteorological record in France) and 1945 [49]. No precise death toll linked to high temperatures has been recorded, while in Italy, excess death has been partly linked to cholera and possibly other diseases, which hot temperatures may have also favoured. Between 20 and 31 July 1987, a heatwave resulted in more than 700 excess deaths in Athens and its vicinity [50]. The 2003 heatwave was a turning point, with heatwaves across Europe resulting in the excess death of about 30,000 persons, with the highest toll in France and Mediterranean countries [51]. The summer 2022 heatwave resulted in more than 61.000 excess deaths across Europe [52]. Supplementary deaths were mainly registered in Spain, Italy, the Balkans, and Greece. The temperature deviation of 1 to 2 degrees in Germany and Scandinavia, with lower reference temperatures, resulted in less harsh absolute temperatures and a much lower resulting caseload.

5.2 Energy and watershed

Higher temperatures result in higher evaporation and evapotranspiration and in warmer water bodies and rivers. Consequently, Rain patterns are disrupted, resulting in drought and flood events. The water from the ice cap melt increases the amount of water available in rivers flowing down from high mountains. However, the snow cover's progressive reduction, with strong interannual variation, jeopardises river flow over the summer.

Heated-up water in rivers increases the risk of algae blossoming and eutrophication. This evolving situation with reduced and warmer water availability raises concerns regarding thermal and nuclear plants requiring cooling. During the 2003 heatwave, some nuclear plants had to be put on hold. In France, an exemption to use water for cooling processes was granted to six nuclear and some thermic plants, even though the rejected water temperature bypassed environmental limits [51]. As people turn to air conditioning and production faces increasing constraints, the soaring electricity demand poses a challenge that will become more acute in a warming climate.

5.3 Impact on trees and forestry

While a historical perspective is lacking, there is no consensus regarding the degree of forest vulnerability [53]. However, this does not mean that the trend is not assessed as worrying. Different parameters can be assessed to measure the plants' stress levels. The Vapour Pressure Deficit (VPD) measures the difference between the air's water pressure and the plant's vapour pressure, triggering evapotranspiration. The relative humidity the air can withhold rises with temperature. By approximation, the VPD is often considered to be the ratio of the air's relative humidity to the theoretical maximum humidity at air saturation for a given temperature. The drier the air, the stronger the aspiration pressure through the plant's leaves. Even a slight temperature increase creates a multiplicative elevation of VPD [54]. Besides, the quantity of water the plant can transport depends on the soil moisture, the size of the trunk, and the leaf area. The plant can regulate the flow by opening or closing the stoma in its leaf cells to different degrees, but under extreme conditions, this remains insufficient to regulate the flow in proportion to water intake and water loss.

The soil moisture at the root level provides information regarding water availability. The soil structure and its water content, jointly with the stand basal area and the root depth, define the volume of water available for a plant. Relative Extractable Water (REW) refers to the stock of water available in the soil, potentially available according to the soil structure. Thereafter, the Climatic Water Demand (CWD), expressed in pressure (kilo Pascal), quantifies the annual evaporative demand that exceeds available water. The Tree Water Deficit (TWD) is calculated at the plant's level to measure the tree's water transportation capacity [55]. It is based on the difference between the past highest stem radial record and the stem radial reading at a given time. Discrepancies between the TWD and the tree's growth reduce the tree's capacity to face a heatwave. The assessment of those factors, combined with meteorological data, satellite remote sensing imagery, and field data, allows us to fine-tune our understanding of local and global dynamics [56]. A longer-term perspective is required regarding land cover and tree die-off phenomena, as there is no direct link between a heatwave or drought event and the observed tree deaths. Evidence of delayed response following heatwaves has been documented since 2003 [57].

Trees' adaptative mechanisms to stress and drought [58] are not at no cost to the plants. Those adaptative mechanisms include evapotranspiration regulation through stoma opening reduction, leaf area reduction, and physiological metabolism slow-down. Stoma closure under heat increases the risk of 'hydraulic failure' and 'carbon starvation,' which affect the plant's metabolism and, thereafter, its capacity to cope with external aggressions. In the same way, a reduction in leaf area can substantially reduce the trees' capacity to store carbohydrates. Thus, the plant's capacity to recover after the shock is reduced [59], which ultimately can result in higher tree mortality. Increased CO2 availability in the atmosphere ceases to be an advantage for plant growth under high temperatures and drought conditions [60]. Photosynthesis is strongly reduced or stopped while the plant closes its stoma under heat stress. Besides, respiration, mainly at night, has a non-linear response to high temperatures [61]. This means that the physiological mechanisms linked to respiration deteriorate considerably or stop functioning beyond a temperature threshold. This, in turn, affects the plants' capacity to recover and run other physiological mechanisms during the day, including photosynthesis.

Heatwaves increase water demand, resulting in drought when the Climatic Water Demand (CWD) exceeds water availability. The trees' adaptative mechanisms, even to mild drought, weaken the plant's resilience capacity. The plant's capacity to cope when successive drought events occur, or other processes like pests or storms, is thus reduced step by step. Ultimately, this can lead to a die-off process. Multi-layers forests allow creating conditions for higher humidity retention, moderating temperatures under the canopy. The vegetation density can be an advantage in that regard. Higher die-off has been recorded in less dense stems per surface area [62]. Besides, a lower leaf area to sapwood (external wood rings allowing water and nutrient transportation) ratio reduces the risk of overcoming the Climatic Water Demand of the plant. This favours younger plants in comparison to old trees with broad leaf areas. As an adaptation strategy, some trees multiply young shoots. Exposure to high temperatures results in slower forest growth [63] and tree die-offs. Larger trees are at higher risk of dying [53].

Quick changes, like alternate precipitation extremes and heatwaves from 1 year to another or within a year, increase trees' vulnerability [64]. Besides, there is an upper limit to the severity and frequency of shocks a tree can withstand, increasing mortality events over time [65]. A humid spring can result in increased leaf development, allowing higher evapotranspiration, but difficult to sustain during heatwaves in the summer period, as stems' radial growth is reduced. This results in leaf loss and reduced plants' resilience capacity [57]. Crown die-offs are most prevalent around rocky outcrops and in soils with poor water retention capacity. The probability of tree death diminishes with soil depth and stand basal area but increases with the trunk diameter at human breast height. In Central Europe, this process has led to an extensive ongoing dieback process [64]. However, the magnitude of the change cannot yet be evaluated. Predicting the future vegetation change and feedback to global climate is challenging. Diebacks have been reported on Norway spruce caused by different factors, possibly drought-induced on weakened trees [66]. European Beech (*Fagus Sylvatica*) dieback has been observed across Central Europe in spring and summer 2019. The 2018 drought following the 2003 drought is believed to have played a major role in the dieback of weakened trees [67].

Finally, the combination of high temperatures and dryer vegetation considerably increases the risk of wildfires. During the 2003 heatwave, more than 25,000 fires were recorded across Portugal, Spain, Italy, France, Austria, Finland, Denmark, and Ireland. In total, about 647,069 hectares were destroyed by fires across Europe. Between the reduction in evapotranspiration linked to high temperatures and wildfires, forest carbon sequestration is endangered. However, the primary role of forests in local climate mitigation remains crucial while considering those challenges.

5.4 Impact on agriculture

The 2003 heatwave has resulted in an estimated drop of 30% in primary productivity [60]. The heatwave began in early June, resulting in early crop development and ripening. However, further in July, the high temperatures, with high evapotranspiration, resulted in a water deficit affecting grain development [68]. Beyond the water availability for rain-fed field crops, like wheat, high temperatures over long periods, including at night, can also affect the plant's physiological limits, affecting yields. The optimum temperature for wheat growth at the heading, anthesis, and grain filling duration is 16 ± 2.3, 23 ± 1.75, and 26 ± 1.53°C, respectively [69]. A high temperature above 40°C inhibits photosynthesis, reducing vegetative growth to the 'zero level.' Tropical night, where temperature does not drop below 20°C, also affects the plant's physiological mechanisms. Over all of Europe, the main sectors hit by the extreme climate conditions were the green fodder supply, the arable sector, potatoes, the livestock sector, and forestry [69]. The fodder deficit varied from 30% (Germany, Austria, and Spain) to 40% (Italy) and 60% in France. The fall in cereal production in the European Union reached more than 23 million tonnes as compared to 2002. Herders have been suffering from rising fodder prices. The 2018 heatwave affected an area 50% larger than in 2003. Pastures and arable land have been especially affected [70]. In 2018, gains in Maize yields were estimated to be 10% in Romania and Hungary, but losses of 10% occurred across Germany and Belgium, resulting in an average loss of 6% for Europe. Even though forest mortality and growth decline are occurring with delay, according to the latest trend, Germany's forestry estimated loss amounted to 105 million cubic metres in 2018.

Since the end of the Second World War, yields have considerably increased with the joint optimisation of varieties and inputs. A more detailed assessment of the quantitative gain compared to the quality of the considered crop varieties would be required to draw precise conclusions. Nevertheless, the production gain has been real. Thus, the gain in productivity over the past period might blur the assessment of counterproductive effects linked to drought and heatwaves. Besides, multiple factors can also affect the crop development. Nevertheless, a study comparing the occurrence of Extreme Weather Disasters (EWD), droughts or heatwaves, and crop production anomalies over the 1964–1990 and the 1991–2015 period concluded that in the second period, production losses in Europe saw an average cumulative increase of 3% per year [71]. Heatwaves have been assessed as being twice as damaging for cereals than for other crops.

Higher temperatures during the crop's vegetative development and even more at critical development stages, like heading, anthesis, and grain filling for cereals, increase the risk of stepping over the plant's physiological limit. While global warming progresses, agriculture will likely face an increasing challenge in maintaining yields. Where, in the past, the main limiting factor was considered to be drought, high temperatures will require much more attention in the forthcoming years. Even if stable in frequency, the disruptions of the North Pole vortex cold events are likely to affect crops that will bud earlier under warming conditions. Abnormal warm conditions followed by cold events could become a more common phenomenon disrupting field crops and overall ecosystems in spring. Nevertheless, shifts in the cultivation calendar may improve the resilience of annual crops. Cooling through spraying water and drip irrigation for high-added-value crops can also mitigate the effect of heatwaves. However, drip irrigation in open fields will face limits when the temperature rises beyond the plant's physiological limit, while spray irrigation will be confronted with reduced water availability in summer. Genetic modification can offer some scope to maximise plants' resistance to different stress factors. However, this potential does not mean the absence of limits. Plants fixing carbon from the atmosphere on a four-carbon chain (C4) have better resilience to high temperatures than plants fixing carbon on a three-carbon chain (C3) [72]. This is because the C4 plants create a stronger partial pressure gradient at the leaf level, which allows for the absorption of more CO_2 from the atmosphere with a smaller opening of the stoma. Gas exchanges between the leaf cells and the air can thus take place with less water loss from evapotranspiration. However, there is no scope for transforming C3 into C4 plants. Besides, C4 crops like maize [73] and sorghum also require considerable amounts of water and have upper limits to heat tolerance, about 35°C, a threshold above which vegetative growth is stopped. Beyond the possible improvements, by specialising excessively plants on limited genetic criteria, the genetic intra-specific variety among individuals is strongly reduced, affecting their capacity to face a changing environment. Increased resistance to drought or slightly higher temperatures might be done at the cost of lower resistance to other aggressions, which may occur. The advantage of annual crops is that the sowing season can be changed annually, while perennial crops or trees with an even longer lifespan require more anticipation. Global warming will increasingly affect European countries from the South to the North. According to prevailing local conditions, ecosystems, and water availability, permanent adaptation to changing conditions will thus be required.

6. GHG emission trend and heatwave occurrence

6.1 GHG emissions and global warming scenarios

Since the first world conference on climate change was held in Rio in 1992 and, more recently, since the Paris Declaration in 2015, GHG emissions have increased year after year. In 2023, with 37.4 Giga tonnes (Gt), emissions increased further by 1.1% compared to 2022.

Far from falling rapidly—as is required to meet the global climate goals set out in the Paris Agreement—CO_2 emissions reached a new record high of 37.4 Gt in 2023 [74]. Atmospheric CO_2 concentration reached 419.3 parts per million (ppm), 51% above pre-industrial level (around 278 ppm in 1750). In practice, the countries' pledges presented in Nationally Determined Contributions (NDC) are declarations of intentions that are not likely to be met and, in most cases, only consider a reduction of emissions in proportion to a growth scenario [75]. Therefore, the agreed effort is to slow the increase rather than decrease emissions. Representative Concentration Pathways (RCP) are elaborated according to different assumptions. Shared Socio-economic Pathways (SSP) have been defined for each RCP, defining possible societal choices influencing the RCP scenario, including adaptation plans. SSP5/RCP 8.5 has been developed as a no-action reference scenario. It considers a continuous increase in GHG emissions without change, the 'business as usual' scenario. In 2024, it seems to be the scenario most closely reflecting the reality of the pathway as it happens.

Comparisons between projections in energy production modelling scenarios and RCP 8.5 present a 90–98% convergence for CO_2 and CH4 emissions [76]. RCP 8.5 has been the most consistent model in coherence with observed historical trends in CO_2 emissions (within 1% for the 2005–2020 period); it also remains a plausible path until 2100 under current policy perspectives [77].

6.2 Heatwave occurrence in Europe according to models

For France, on average, in a scenario with +4°C at the end of the century, the average annual number of heatwave days would be eight to ten times higher in 2100 in comparison to the 1976–2005 reference period [77]. On the French Mediterranean coast, the increase would be eleven to twelve times. Under RCP 8.5, by 2071–2100, heatwaves could start to occur in early May until mid-October [78]. The heatwave episodes are expected to become longer, up to 2 months in the summer period, with temperatures up to 6°C above the maximum observed over the 1976–2005 reference period.

According to the SSP5/RCP.8.5 model, in France, the average heating by 2080–2099 compared to the 1995–2014 period would be 3–4°C on average over the winter period and above 6°C between July and September [11]. For Germany, the projections show an increase of 4.5 to 6°C between June and October, with a rise above 5°C from July to September. For Germany and Poland, the average maximal temperature for the 1995–2014 reference period was 12°C. For Sweden, the increase in average maximal summer temperature is similar to the one foreseen for Germany and Poland, but starting from substantially lower reference temperatures, with an average maximal temperature of 6°C for the 1995–2014 reference period [11].

Temperature variation does not have the same meaning everywhere, but it will result in different challenges depending on the geographic area considered. In Sweden, the increase in summer temperature will affect the snow cover and glaciers, while changing rain patterns might additionally increase the risk of droughts. High seasonal temperatures and droughts could thus endanger forests. In Southern Europe, high seasonal temperatures will more rapidly become a threat to human health and production systems while jeopardising ecosystems.

7. Conclusion

We lack temporal distance to evaluate the possible disruption and changes in energy fluxes at the earth's surface. However, the acceleration of heatwave events in frequency and intensity since the 1990s and even more into the twenty-first century underlines the effect of global warming. Meteorological phenomena can worsen or attenuate the trend, but they do not deeply change the dynamic humanity and all living beings face. Heatwaves beyond a temperature threshold, which varies among crops and varieties, stop plant growth, jeopardise yields at critical development stages, and endanger ecosystems. Such events will also increasingly affect territories in Southern Europe and progressively in Northern countries. Besides, recurrent heatwaves with increasing intensity endanger trees and forests all over Europe. This could put the vegetation at risk and reduce the attenuation effect. Mediterranean and Central European continental areas will be exposed first. Over the twenty-first century, sustaining vegetation permanent land cover will be a major challenge, while its reduction will affect the capacity to mitigate local climate.

However, land cover remains a key element among attenuation mechanisms that can be influenced. Forest cover influences the local climate, reduces temperature, and favours more regular rain patterns. Vegetation's role may thus not be reduced to carbon sequestration. Sufficient interspecific and intra-specific biodiversity is important to maximise the resilience of those ecosystems. However, urban sprawl and, in some cases, extension of annual crops or the change of land destination with solar panel fields instead of natural areas affect the local albedo and evapotranspiration. Adaptation mechanisms can also help increase ecosystems and crop resilience locally. The sowing period can be adjusted, some improvements in varieties can be made, and irrigation techniques can be applied, among other mechanisms. However, those solutions also have weaknesses, affecting production systems' resilience and implying choices in water allocation. In the short term, annual crops have more scope for adaptation, while perennial crops or trees with an even longer lifespan might require more anticipation. Northern European countries have a bit more time, even though the 2003 and 2018 heatwaves showed they need to be prepared for the upcoming change. In any case, adaptation measures need to be defined on a case-by-case basis. However, reducing greenhouse gases should be a priority, requiring decoupling economic growth scenarios from energy consumption. The increasing demand for air cooling devices in summer will be challenging in this regard. Besides, at the local level, territorial planning requires strategic insight to attenuate, as far as possible, the effects of global warming.

Further research is required to assess the local impact of global warming on living systems and possible attenuation measures at a territorial level. This research should associate meteorological data, a network of forest tree observations measuring soil moisture, VPD, and die-out count, and broader satellite imagery soil moisture data. Assessing land cover dynamics, LAI, and crop yields could add relevant complementary insight. The work to be conducted in the South of France and possibly in neighbouring Mediterranean countries could provide useful conclusions to allow more Northern European countries to anticipate forthcoming changes.

Acknowledgements

The present publication is based on the work of the author, bibliographic references, and experience drawn from different parts of the world.

Conflict of interest

The author declares no conflict of interest.

Appendices and nomenclature

AMOC	Atlantic Meridional Overturning Circulation
CO_2	carbon dioxide
C3	plants absorbing carbon from the atmosphere on a three-carbon chain
C4	plants absorbing carbon from the atmosphere on a four-carbon chain
CWD	climatic water demand
EWD	extreme weather disasters

GHG	greenhouse gas
IPCC	intergovernmental panel on climate change
LAI	leaf area index—measuring leaf surface available for evapotranspiration
LULUCF	land use, land use change, and forestry
NDC	nationally determined contributions
PACA	Provence-Alpes-Côte d'Azur (Region—France)
RCP	representative concentration pathway
REW	relative extractable water
SSP	shared socio-economic Pathways
TWD	tree water deficit
UK	United Kingdom
VPD	vapour pressure deficit

Author details

Julian Schlubach
Independent Expert, Aix en Provence, France

*Address all correspondence to: julian.schlubach@gmail.com

References

[1] European Environment Agency – Extreme Weather: Floods, Droughts and Heatwaves. 2024. Available from: https://www.eea.europa.eu/en/topics/in-depth/extreme-weather-floods-droughts-and-heatwaves

[2] European State of the Climate – 2023. Available from: https://climate.copernicus.eu/esotc/2023

[3] William JF et al. Machine learning identifies ecological selectivity patterns across the end-Permian mass extinction. Paleobiology. 2022;**1**:12-13. DOI: 10.1017/pab.2022.1

[4] University of Hamburg. Three Critical Factors in the End-Permian Mass Extinction. Germany: University of Hamburg, Science Daily; 2022. Available from: www.sciencedaily.com/releases/2022/03/220301131200.htm

[5] Shapiro AV et al. Solar-cycle irradiance variations over the last four billion years. Astronomy & Astrophysics. 2020;**636**:1-2. Open Access article, published by EDP Sciences. DOI: 10.1051/0004-6361/201937128

[6] World Meteorological Organisation – Heatwaves. Available from: https://wmo.int/topics/heatwave

[7] Meteo France. Vagues de chaleur et changement climatique. 2023. Available from: https://meteofrance.com/changement-climatique/observer/changement-climatique-et-vagues-dechaleur

[8] Pongratz J. Land use Efects on climate: Current state, recent Progress, and emerging topics. In: Vegetation and Climate Change. Vol. 7. Current Climate Change Reports. Germany: Springer; 2021. pp. 99-120. Available from: https://link.springer.com/article/10.1007/s40641-021-00178-y

[9] Schlubach J. Downscaling model in agriculture in Western Uzbekistan climatic trends and growth potential along field crops physiological tolerance to low and high temperatures. Heliyon. 2021;**7**(5):e07028. Available from: https://authors.elsevier.com/sd/article/S2405844021011312

[10] Schlubach J. Crops and ecosystem services, a close interlinkage at the Interface of adaptation and mitigation, chapter 7. In: Kumar V, editor. Global Warming - a Concerning Component of Climate Change. London, UK: IntechOpen Limited; 2024. DOI: 10.5772/intechopen.1000331

[11] World Bank Climate Change, Open Portal. Available from: https://climateknowledgeportal.worldbank.org

[12] Bader J, et al. Global Temperature Modes Shed Light on the Holocene Temperature Conundrum. Available from: https://www.nature.com/articles/s41467-020-18478-6

[13] Orth R et al. Did European temperatures in 1540 exceed present-day records? Environmental Research Letters. 2016;**11**:114021 Available from: https://iopscience.iop.org/article/10.1088/1748-9326/11/11/114021

[14] Murden S. The heatwave of July 1808. In: Weather in Georgian England. Wordpress; 2018. Available from: https://georgianera.wordpress.com/2018/07/17/the-heatwave-of-1808/

[15] Cohen J. Heatwaves Throughout History - A Look Back at Some of the

most Infamous and Deadly Heatwaves in History. 2023. Available from: https://www.history.com/news/heat-waves-throughout-history

[16] Büntgen U, Brázdil R, Heussner K-U, Tegel W. Combined dendro-documentary evidence of central European hydroclimatic springtime extremes over the last millennium. Quaternary Science Reviews. 2011;**30**:27-28. DOI: 10.1016/j.quascirev.2011.10.010. Available from: https://www.researchgate.net/publication/228469464_Combined_dendro-documentary_evidence_of_Central_European_hydroclimatic_springtime_extremes_over_the_last_millennium

[17] Waybackmachine 1900-1949. Available from: https://web.archive.org/web/20140508223425/http://booty.org.uk/booty.weather/climate/1900_1949.htm

[18] Available from: https://www.meteo-paris.com/chronique/annee/1947

[19] Historical Record of Bucharest, Romania Weather for the Year 1952 based on NOAA Data – Bucharest Weather in 1952. Available from: https://www.extremeweatherwatch.com/cities/bucharest/year-1952

[20] Metmatters; Recalling the 1976 drought: 40 years on. 2016. Available from: https://www.rmets.org/metmatters/recalling-1976-drought-40-years

[21] Butturini P. United Press International (UPI). European Heatwave Breaking Records. 1983. Available from: https://www.upi.com/Archives/1983/07/30/European-heat-wave-breaking-records/6550428385600/

[22] Chrysopoulos P. Extremely High Temperatures in Greece Brings Back Memories of 1987 Killer Heatwave. Greek Reporter; 2017. Available from: https://greekreporter.com/2017/06/29/extremely-high-temperatures-in-greece-brings-back-memories-of-1987-killer-heatwave/

[23] Met office. Hot Spell. 1990. Available from: https://www.metoffice.gov.uk/binaries/content/assets/metofficegovuk/pdf/weather/learn-about/uk-past-events/interesting/1990/hot-spell-august-1990---met-office.pdf

[24] Institut National de l'Audiovisuelle - Nantes en 1990: une canicule qui posait des questions. Available from: https://www.ina.fr/ina-eclaire-actu/nantes-en-1990-une-canicule-qui-posait-des-questions

[25] Laurynas Klimaviˇcius, Egidijus Rimkus; Institute of Geosciences, Vilnius University, Vilnius, Lithuania. Compound drought and heatwave events in the eastern part of the Baltic Sea region. 2023. Available from: https://www.sciencedirect.com/science/article/pii/S0078323423000684

[26] Karl TR, Knight RW, Bake B. The record-breaking global temperatures of 1997 and 1998: Evidence for an increase in the rate of global warming? Geophysical Research Letters. 2000;**27**(5):719-722. Available from: https://agupubs.onlinelibrary.wiley.com/doi/epdf/10.1029/1999GL010877

[27] Meteorological office. Record High Temperatures Verified. UK: Press Office; 2022. Available from: https://www.metoffice.gov.uk/about-us/news-and-media/media-centre/weather-and-climate-news/2022/record-high-temperatures-verified

[28] Météo France. Vagues de chaleur et changement climatique. 2023. Available from: https://meteofrance.

com/changement-climatique/observer/changement-climatique-et-vagues-de-chaleur#:~:text=Les%20vagues%20de%20chaleur%20recens%C3%A9es,eu%20lieu%20en%20ao%C3%BBt%202022

[29] ECMWF Newsletter No. 99 – Autumn/Winter 2003 - The Exceptional Warm Anomalies of Summer. 2003. Available from: https://www.ecmwf.int/sites/default/files/elibrary/2003/14625-newsletter-no99-autumnwinter-2003.pdf

[30] Struzewska J, Kaminski JW. Formation and Transport of Photooxidants over Europe During the July 2006 Heatwave. 2007. Available from: https://acp.copernicus.org/preprints/7/10467/2007/acpd-7-10467-2007.pdf

[31] Thomas R. "Vague de chaleur dévastatrice en Europe du Sud". Le Figaro (in French). 2017. Available from: https://www.lefigaro.fr/international/2007/07/26/01003-20070726ARTFIG90265-vague_de_chaleur_devastatrice_en_europe_du_sud.php

[32] Barriopedro D, Fischer EM, Luterbacher J, Trigo RM, Garcia-Herrera R. The hot summer of 2010: Redrawing the temperature record map of Europe. Science. 2011;**332**(6026):220-224. Available from: https://www.science.org/doi/10.1126/science.1201224

[33] MeteoLux - Forte chaleur au Luxembourg - Record de la température maximale pour le mois de juillet. 2015. Available from: https://www.meteolux.lu/fr/actualites/forte-chaleur-au-luxembourg-record-de-la-temperature-maximale-pour-le-mois-de/?lang=fr

[34] Copernicus; European State of the Climate 2017, Climate in 2017 - Focus Region: Southwest Europe. Available from: https://climate.copernicus.eu/climate-2017-focus-region-southwest-europe

[35] The Extremely Hot and Dry 2018 Summer in Central and Northern Europe from a Multi-Faceted Weather and Climate Perspective; Copernicus, Highlight Paper. 2023. Available from: https://nhess.copernicus.org/articles/23/1699/2023/

[36] Copernicus - Heatwaves Grip Parts of Europe, Asia and North America in the First Half of 2022. 2022. Available from: https://climate.copernicus.eu/heatwaves-grip-parts-europe-asia-and-north-america-first-half-2022

[37] The Guardian. UK Weather Returns to Normal after Record-Breaking Heatwave. London: The Guardian; 2011. Available from: https://www.theguardian.com/uk/2011/oct/03/uk-weather-normal-record-heatwave?newsfeed=true

[38] Met Office. July 2013 Heatwave. Available from: https://www.metoffice.gov.uk/weather/learn-about/weather/case-studies/heat-wave-july2013

[39] Met Office. Hot Spell June 2017. Available from: https://www.metoffice.gov.uk/binaries/content/assets/metofficegovuk/pdf/weather/learn-about/uk-past-events/interesting/2017/hot-spell-june-2017---met-office.pdf

[40] World Meteorological Organisation. Europe Experiences Widespread Flooding and Severe Heatwaves in 2023. Geneva, Switzerland: World Meteorological Organisation; 2024. Available from: https://wmo.int/news/media-centre/europe-experiences-widespread-flooding-and-severe-heatwaves-2023

[41] Loeb NG et al. CERES top-of-atmosphere earth radiation budget climate data record: Accounting for in-orbit changes in instrument calibration. Remote Sensing. 2016;**8**(3):182. Available from: https://www.mdpi.com/2072-4292/8/3/182

[42] Rousi E, Kornhuber K, Beobide-Arsuaga G, et al. Accelerated western European heatwave trends linked to more-persistent double jets over Eurasia. Nature Communications. 2022;**13**:3851. DOI: 10.1038/s41467-022-31432-y

[43] Seneviratne SI, Corti T, Davin EL, Hirschi M, Jaeger EB, Lehner I, et al. Investigating soil moisture–climate interactions in a changing climate: A review; Institute for Atmospheric and Climate Science. Earth-Science Reviews. 2010;**99**:125-161. Available from: https://www.sciencedirect.com/science/article/abs/pii/S0012825210000139?via%3Dihub

[44] Stegehuis AI, Vogel MM, Vautard R, Ciais P, Teuling AJ, Seneviratne SI. Early summer soil moisture contribution to Western European summer warming. Journal of Geophysical Research: Atmospheres. 2021;**126**:e2021JD034646. DOI: 10.1029/2021JD034646

[45] Vautard R, Yiou P, D'Andrea F, de Noblet N, Viovy N, Cassou C, et al. Summertime European heat and drought waves induced by wintertime Mediterranean rainfall deficit. Geophysical Research Letters. 2007;**34**:L07711. DOI: 10.1029/2006GL028001. Available from: https://agupubs.onlinelibrary.wiley.com/doi/10.1029/2006GL028001

[46] Seneviratne S, Lüthi D, Litschi M, et al. Land–atmosphere coupling and climate change in Europe. Nature. 2006;**443**:205-209. DOI: 10.1038/nature05095

[47] Joly D et al. Les types de climats en France, une construction spatiale. Cybergeo: European Journal of Geography [online], Cartographie, Imagerie, SIG. 2010;**501**:12-13. Available from: http://cybergeo.revues.org/index23155.html

[48] Geo Confluence - Gulf Stream, Circulation de Retournement, AMOC. 2021. Available from: https://geoconfluences.ens-lyon.fr/glossaire/gulf-stream-circulation-de-retournement-amoc

[49] Pozzi L, Ramiro Fariñas D. The heat-wave of 1911. A largely ignored trend reversal in the Italian and Spanish transition? Annales de démographie historique, Cairn info. 2010;**2**(120):147-178. Available from: https://www.cairn.info/revue-annales-de-demographie-historique-2010-2-page-147.htm

[50] The Guardian – Archive. 1987: Hundreds Die in Greek Heatwave. Available from: https://www.theguardian.com/world/2022/jul/13/hundreds-die-in-greek-heatwave-archive-1987

[51] UNEP. Impacts of Summer 2003 Heatwave in Europe. Environment Alert Bulletin. 2003. Available from: https://www.unisdr.org/files/1145_ewheatwave.en.pdf

[52] Treisman R. Heatwaves in Europe Killed more than 61,600 People Last Summer, A Study Estimates. 2023. Available from: https://www.npr.org/2023/07/12/1187068731/heat-waves-europe-deaths-study

[53] Allen CD, Breshears DD, McDowell NG. On underestimation of global vulnerability to tree mortality and

forest die-off from hotter drought in the Anthropocene. Ecosphere. 2015;**6**(8):129. DOI: 10.1890/ES15-00203.1

[54] Eamus D, Boulain N, Cleverly J, Breshears DD. Global change-type drought-induced tree mortality: Vapor pressure deficit is more important than temperature per se in causing decline in tree health. Ecology and Evolution. 2013;**3**(8):2711-2729. DOI: 10.1002/ece3.664

[55] Salomón RL, Peters RL, Zweifel R, et al. The 2018 European heatwave led to stem dehydration but not to consistent growth reductions in forests. Nature Communications. 2022;**13**:28. DOI: 10.1038/s41467-021-27579-9

[56] Small EE et al. Estimating soil evaporation using drying rates determined from satellite-based soil moisture records. Remote Sensing. 2018;**10**(12):1945. DOI: 10.3390/rs10121945

[57] Schuldta B, Burasb A, Arend M, et al. A first assessment of the impact of the extreme 2018 summer drought on central European forests. Opinion paper. In: Basic and Applied Ecology. Germany: Elsevier; 2018. Available from: https://www.sciencedirect.com/science/article/pii/S1439179120300414

[58] Klein T, Hoch G. Tree carbon allocation dynamics determined using a carbon mass balance approach. New Phytologist. 2015;**205**:147-159. DOI: 10.1111/nph.12993. Available from: https://nph.onlinelibrary.wiley.com/doi/epdf/10.1111/nph.12993

[59] Dorman M, Svoray T, Perevolotsky A, Sarris D. Forest performance during two consecutive drought periods: Diverging long-term trends and short-term responses along a climatic gradient. In: Forest, Ecology and Management. Elsevier; 2013. Available from: https://citeseerx.ist.psu.edu/document?repid=rep1&type=pdf&doi=7d2cfc7db40d20b5d1345cbe151d3227e1051e20

[60] Ciais P et al. Europe-wide reduction in primary productivity caused by the heat and drought in 2003. Nature. 2005;**437**(7058):529-533. Available from: https://www.nature.com/articles/nature03972

[61] Atkin OK, Tjoelker MG. Thermal acclimation and the dynamic response of plant respiration to temperature. Trends in Plant Science. 2005;**8**:343-351. Available from: https://www.sciencedirect.com/science/article/abs/pii/S1360138503001365

[62] Steel EJ, Fontaine JB, Ruthrof KX, Burgess TI, Hardy GESTJ. Changes in structure of over- and midstory tree species in a Mediterranean-type forest after an extreme drought-associated heatwave. Austral Ecology. 2019;**44**:1438-1450. DOI: 10.1111/aec.12818

[63] Zhao M, Running SW. Drought-induced reduction in global terrestrial net primary production from 2000 through 2009. Environmental Science. 2010;**329**(5994):940-943. DOI: 10.1126/science.1192666. Available from: http://files.ntsg.umt.edu/data/NTSG_Products/MOD17/Zhao_Running_Science_2010.pdf

[64] Hammond WM. Global field observations of tree die-off reveal hotter-drought fingerprint for Earth's forests. Nature Communications. 2022;**13**:1761. Available from: https://www.nature.com/articles/s41467-022-29289-2

[65] Marqués L, Ogle K, Peltier DMP, Camarero JJ. Altered climate memory characterizes tree growth during forest dieback. Elsevier. Agricultural and Forest

Meteorology. 2022;**314**:108787. DOI: 10.1016/j.agrformet.2021.108787

[66] Godbold DL. Dieback in Norway Spruce - Causality and Future Management Strategies. 2013. Available from: https://forschung.boku.ac.at/en/projects/8952

[67] Rukh S et al. Distinct responses of European beech (Fagus sylvatica L.) to drought intensity and length—A review of the impacts of the 2003 and 2018-2019 drought events in Central Europe. Forests. 2023;**14**:248. Available from: https://www.mdpi.com/1999-4907/14/2/248

[68] COPA. COGECA - Assessment of the Impact of the Heatwave and Drought of the Summer 2003 on Agriculture and Forestry. COPA; 2004 Available from: http://docs.gip-ecofor.org/libre/COPA_COGECA_2004.pdf

[69] Khan A, Ahmad M, Ahmed M, Iftikhar Hussain M. Rising atmospheric temperature impact on wheat and Thermotolerance strategies. Plants. 2021;**10**:43. DOI: 10.3390/plants10010043

[70] Beillouin D et al. Impact of extreme weather conditions on European crop production in 2018. Philosophical Transactions of the Royal Society. 2020;**B 375**:20190510. DOI: 10.1098/rstb.2019.0510

[71] Brás TA et al. Severity of drought and heatwave crop losses tripled over the last five decades in Europe. Environmental Research Letters. 2021;**16**:065012 DOI 10.1088/1748-9326/abf004. Available from: https://iopscience.iop.org/article/10.1088/1748-9326/abf004/meta

[72] Yin X, Struik PC. C3 and C4 photosynthesis models: An overview from the perspective of crop modelling. NJAS - Wageningen Journal of Life Sciences. 2009;**57**:27-38

[73] Rahman et al. Evaluation of Maize Hybrids for Tolerance to High Temperature Stress in Central Punjab. United States: Columbia International Publishing; 2013

[74] IEA. CO_2 Emissions in 2023. Paris: IEA; 2024. Available from: https://www.iea.org/reports/co2-emissions-in-2023

[75] Schlubach J. Carbon trajectory and adaptation plans. International Journal on Agriculture Research and Environmental Sciences. 2024;**5**(2):1-2. DOI: 10.51626/ijares.2024.05.00044. Available from: https://skeenapublishers.com/journal/ijares/IJARES-05-00044.pdf

[76] Hausfather Z. Explainer: The High-Emissions 'RCP8.5' Global Warming Scenario. Carbon Brief; 2019. Available from: https://www.carbonbrief.org/explainer-the-high-emissions-rcp8-5-global-warming-scenario/

[77] Schwalm CR, Glendon S, Duffy PB. In: Dickinson RE, editor. RCP8.5 Tracks Cumulative CO_2 Emissions. Los Angeles, CA: University of California; 2020. Available from: https://www.pnas.org/doi/10.1073/pnas.2007117117

[78] Schneider M, Corre L. Evolution passée et future des vagues de chaleur en France. Meteo France; 2021. Available from: http://www.meteo.fr/cic/meetings/2022/aic/presentations/obs_model_schneider.pdf

Chapter 4

Why Pricing Carbon is Still a Good Idea

Gary Yohe

Abstract

In the decades-long debate about how best to design policies that would most efficiently and equitably limit the emission of carbon dioxide (CO_2) from all human sources, economists have persistently called for designs that included some way of pricing carbon. Nonetheless, but with some notable exceptions, very few implemented policies have moved very far away from traditional command and control strategies wherein the only price is what economists call a "shadow price" that is barely visible to the general public. To understand why that is so despite economists' objections and to keep those reasons front of mind, it is useful to review the underlying economics from time to time, especially in the context of pricing's advantages in coping with the enormous uncertainties and long time horizons that add unprecedented complexity to the regulatory problem. This essay does just that. It starts with the theory behind economists' unfaltering position, starting with Marty Weitzman's seminal comparison of price and quantity controls under uncertainty in 1974. It continues by considering how the application of Weitzman's finding to environmental controls can handle the twin monumental challenges to design issues posed by coping effectively with uncertainty over long periods of time. A third section provides evidence that some with position in the policy arena have not been shy in speaking these insights to power over many years before concluding remarks offering some suggestions about what the future could hold, depending on the results of the 2024 presidential election in the United States.

Keywords: uncertainty, mitigation policy, price controls, quantity controls, carbon taxes, cap and trade, climate risks

1. Introduction

My voice has been among the many who have long been arguing that a direct tax on carbon would be the best idea for regulating carbon dioxide emissions, but most of us have also conceded that pricing carbon by some other market mechanism like implementing a "cap and trade" program on emissions permits would be fine. Economic theory has taught us that the efficiencies of either would always dramatically reduce the expected cost of society's meeting any of its politically determined and time-dependent emissions target by, instead, implementing rigid, uneven, and cumbersome "command and control" processes. Here, I explore the basis of this assertion before pondering the ultimate "Why" question circling around the entire issue – "If pricing carbon is such a good idea, why is it so difficult to get the US federal government to agree to do something in that regard?"

2. The economic theory that favors a carbon tax for macro-scale climate policies

Martin Weitzmen [1] laid down the fundamental economics of comparing these three options in the *Review of Economic Studies* under the title "Prices versus Quantities"; they come directly from the first principles of economic efficiency. The title should have included "under uncertainty" to convey its most important context. That would make it clear why, in the climate world, his paper can be interpreted a "tax vs cap and trade vs command and control (broadly defined)" question. Yohe [2] both expanded the title and developed a more generalizable context for its application to real policy choices.

To see how the Weitzman comparisons can be derived from first principles of economics, consider three descriptions of how each option could be designed to limit emissions of a pollutant like carbon dioxide to a specific target over a fixed period of time of a decade or three (say, from 2024 until 2035 or even from now until 2050). Assume, as is reasonable, that the target and the timeframe will have been set by a political process operating at nearly any size of government. The three options are:

1. Institute a command-and-control regime that would send strict and rigid quantity and technological regulations to every source of ambient carbon so that the sum of their emissions across all sources would be no larger than the designated target over the specified period of time;

2. Create a permit market structure wherein each source would need a license for each ton of carbon emitted. Initial allocations of the permits for each year could be distributed free of charge by the government and each source could buy more from another source or sell some at the market clearing price. The total number of licenses available for the specified time period would sum to the selected target.

3. Set a price (tax) on all carbon emissions that would likely increase over time so that the sum of *expected* emissions over the specified time period would match the targeted total achieved with certainty by the market scheme described in #2.

Let us make the easiest comparison by first taking explicit account of the behaviors of multiple emitters in the mix as described in Yohe [3]. Option (#1) is always the worst choice because it allows for no cost minimizing flexibility either across sources or over time. It is therefore more expensive to devise and to administer and it is also more difficult to adjust at scale over time. The other two options overcome both of these obstacles by design, and so their expected economic costs are necessarily smaller (and sometimes by quite a bit).

The choice between the other two options depends on their relative expected efficiencies emissions-decisions that sources want to make over time as market conditions for their products change from quarter to quarter or year to year. Option (#2) allows sources that want to produce more output in a particular year to augment their allocations of permits by purchasing permits from other sources who find themselves with a surplus. These sources, perhaps supplying other product markets, want to slow production and accept lower sales for whatever reason. This transaction makes both participants happy – one makes more money because it can sell more products, and the other makes more money because it can sell pieces of paper for which it would otherwise have no immediate value. Most importantly, total emissions across all sources would be fixed, and the permit market would clear every year at an equilibrium.

By contrast, the tax option always allows sources to produce whatever they want without needing a permission slip for the requisite carbon permits. All they have to do is pay a fee for every ton of carbon that they want to emit so that their dynamic output decisions internalize at least some of the external social costs of their emissions. Emissions from every source can vary over time (by more than they would facing a permit market where the price of carbon can be high or low relative to the tax). It follows that total emissions across all sources can vary from 1 year to the next, but the expected sum of cumulative emissions over the targeted time period should match the policy target (assuming that the initial year's tax was set correctly).

The choice between these two equivalent options (on average) therefore depends on whether the flexibility in total annual emissions over time allowed by the tax saves more in reducing the expected cost of restricting total emissions to the target than the extra loss in the expected economic benefits across the targeted time period. Why this extra social cost from variable annual emissions? Depending on the curvature of their social damage functions, the *expected net social benefits* with variable emissions under the tax can be significantly smaller over the time period than the net social benefits of *expected emissions* – a level of net social benefit that can be achieved most efficiently over the selected time period by the fixed total emissions achieved by a cap and trade permit market. Both of these expected net social benefits depend directly upon the variance in annual emissions allowed by the tax.

For sulfur dioxide (acid rain), annual variability in emissions allowed by a tax can be very expensive when dangerous damage thresholds of acidity are crossed from time to time in lakes, across fields and forests, on buildings and cars, and so on, that is, extra damages suffered during high emissions years that exceed those thresholds can be much greater than any of the marginal damages avoided during low emissions years. For carbon, though, annual variability in emissions is essentially costless because of one well-established fact about how the climate system works on an aggregate global level: as reported in NRC [4], climate change impacts generally depend on temperature change that, in turn, depends directly on changes in atmospheric concentrations of gases like carbon dioxide. Concentrations depend fundamentally on *cumulative* emissions from 1850 – the marker of the beginning of the Industrial Revolution. Nearly 200 years later, even current annual emissions that are historically high make only small and nearly constant marginal contributions to historically high concentrations so that the effect of annual variability around those high levels is functionally insignificant when reflected by temperature change.

It therefore follows from economic theory that cap and trade should be preferred for limiting sulfur emissions. For carbon emissions, though, the low expected marginal cost of variability in emissions that have a zero mean over the policy period makes a tax regime the better choice.

3. Solving two important complications: Time and uncertainty

One complication arises from the obvious fact that mitigation plans are made to cover many decades or even a century or two. If atmospheric concentrations are to be stabilized at any point in the future, however, emissions must eventually fall by 80% of current levels. It follows that a carbon tax must increase and permit supplies must fall over time, but from what initial level and how fast?

Thinking about the challenge of these enormous long-term planning horizons now makes it clear that nobody should be expected to set mitigation targets in 2024 for 2100 and beyond. There is just too much uncertainty in our understanding of the climate system and in our understanding of how the global economy will evolve over centuries under any one of many possible geo-political regimes. It therefore makes sense that policy targets should be created for the long term, but policy designs should be broken into manageable chunks of time. That is to say, policymakers must be experts to make mid-course corrections based on a new understanding of what is going on.

IPCC [5] highlighted this reality when it produced a consensus agreement among the members of the United Nations Framework Convention on Climate Change (UNFCCC) that "Responding to climate change involves an iterative risk management process that includes both adaptation and mitigation and takes into account climate change damages, co-benefits, sustainability, equity, and attitudes to risk". The two to three decadal time horizons that framed the descriptions of the options recorded above fit right into that iterative framework, so the second complication is covered.

At first blush, the first issue looks to be a little more challenging, but it was actually solved in Hotelling [6]. The "Hotelling" rule summarizes his finding: to maximize social benefits from exploiting an exhaustible resource over time, add an *initial* "scarcity rent" to the cost of extraction and let it increase over time at the rate of interest. That is to say, a time-dependent "scarcity tax" added to extraction cost was a control variable in his work on a naturally constrained intertemporal optimization problem just like the carbon tax introduced by Nordhaus [7] was a control variable in his seminal analysis of how to solve an unconstrained optimization problem about human contributions to global warming where the object was to maximize net discounted global welfare when carbon emissions from fossil fuel caused worldwide economic damages that would increase over time.

These two contributions to economic theory now confirm that the design of modern policies described above is appropriate even though modern emissions targets are the product of political decisions and not solutions to a welfare optimization problem. The policy goal today is to minimize the expected costs of hitting those societal targets by, in the case of cap and trade, turning cumulative carbon emissions over the design period into an artificially created exhaustible resource problem for emitters and trust that the permit market will clear each year at levels that will approximate an efficient distribution across time. We can expect, therefore, that the permit price will begin at a level that reflects Hotelling's initial scarcity rent and trend up at the market rate of interest. By analogy, then, the equivalent carbon tax should initially be set at a scarcity rent for carbon that sets expected cumulative emissions equal to the target if it, too, increases at the rate of interest. Interestingly enough, that is exactly the trajectory that Nordhaus produced last century when he turned to minimizing the discounted cost of achieving a temperature target.

4. Evidence that the economics was taken seriously by those who speak to power

Given "the significant risks that climate change poses to human society and the environment," members of the US Congress asked the US National Academies of Science, Engineering, and Medicine [8] early in the Obama administration to "provide a strong motivation to move ahead with substantial response efforts"; and to describe those efforts while they were at it. The result was a collection of four-panel assessments and a synthesis report. One of their most significant findings is summarized on the third page of its summary briefing document:

> *"The most effective way to amplify and accelerate current state, local, and private sector efforts, and to minimize overall costs of meeting a national emissions reduction target, is with a comprehensive, nationally-uniform price on CO 2 emissions, with a price trajectory sufficient to drive major investments in energy efficiency and low-carbon technologies. In addition, strategically targeted complementary policies are needed to ensure progress in key areas of opportunity where market failures and institutional barriers can limit the effectiveness of a carbon pricing system".*

Notice that there is no mention of political obstacles, but there is mention of complementary policies and an entire section on "iterative risk management". It closes with:

> *"A valuable framework for making decisions about America's climate choices is iterative risk management. This refers to a process of systematically identifying risks and possible response options, advancing a portfolio of actions that emphasize risk reduction and are robust across a range of possible futures, and revising responses over time to take advantage of new knowledge, information, and technological capabilities."*

Getting NASEM to pick up on pricing carbon and putting it into context is one thing, but seeing actual action is quite another. Happily, there is plenty of evidence that climate change has spread into the political discourse in the United States since the Academy report. For example, the American Clean Energy and Security Act of 2009 [9] H.R. 2454 in 111th Congress named Waxman-Markey for its primary sponsors would have established a variant of an emissions trading mechanism similar to the European Union Emissions Trading System (EU ETS; [10]. It was approved by the House of Representatives by 219-212 in June of 2009, even though there was no chance of overcoming a threatened Republican filibuster in the Senate. It was more ambitious than both the Lieberman-Warner Bill and President Obama's proposal, but it did not have universal support from the environmental community [11].

Much later, the Inflation Reduction Act of 2023 [12] funded 10 years of climate-friendly investment in infrastructure, alternative energy, electric vehicles, and additional charging infrastructure to promote the demand side of the vehicle market. Most notably, for present purposes, it did not include pricing carbon despite some encouragement to do so. Perhaps most importantly, it was complemented during President Biden's first term by an increasing number of programs that look a lot like creative applications of command and control mechanisms:

- The Security and Exchange Commission [13], after years of drafting a review, finally implemented new reporting rules for material risk from climate change and climate policy. They rely on climate savvy investors in financial markets to motivate corporations to become better climate citizens because it is in their own best interest;

- The Federal Reserve System [14] continued the process of characterizing severe global recession scenarios accompanied by a period of heightened stress in both commercial and residential real estate markets (caused by sudden enormous contributions to global sea level rise from accelerated melting and disintegration of Antarctic Ice Sheets). They rely on word of successful strategies to prevent emerging climate risks from spreading quickly across the entire banking system (see [15]);

- The Environmental Protection Agency [16] continued to strengthen a variety of environmental exposure regulations that improve human welfare one threat at a time, with a recent focus on fossil fuel-generating power systems; and

- The Department of Transportation [17] has recently announced a series of higher fleet mileage requirements that will move supply and demand sides of transportation markets toward increased fuel economy and corresponding lower emissions of hazardous pollutants including greenhouse gases, and so on.

The list can be extended, but the point is clear – climate mitigations of many types have been successfully implemented across the United States for more than a decade, notwithstanding persistent intransigence in the legislative and executive branches of the federal government.

5. Some concluding thoughts

Intransigence in Washington does not mean that pricing carbon is a bad idea for federal-level policy. It is not. After all, some nations, like the entire European Union, have been quite successful in pricing carbon. They were the first to take that approach. They are now joined by the United Kingdom, Australia, New Zealand, South Korea, and China, which are on the list of countries whose daily spot prices for carbon emission allowances are listed by CarbonCredits [18]. On May 11, 2024, the EU ETS price was just over $77 per ton of carbon. Even the northeast region of the United States implemented its own limited multistate Regional Greenhouse Gas Initiative (RGGI) in 2009; RGGI [19] reported that the market clearing auction price in March of that year was $16.00 per ton of carbon.

As reported in Macrotrends [20], United States carbon emissions have been on a short downward trend since around 2020. That trend, even after COVID-19 impacts have passed, shows that things have been getting done across the United States to hasten its mitigation efforts. It has not been enough, though, for any number of reasons that do not really matter for this paper. More has to be done, and pricing carbon is still the cheapest way to do that.

Americans have been allergic to adding new visible taxes since the Boston Tea Party in revolutionary days, but that seems to have been changing recently across much of the population with regard to carbon emissions. As reported from public surveys from late in 2023 by Yale Climate Communication [21], 72% of the US population understands that global warming is real, and 68% think that fossil fuel companies should pay a tax on the carbon that they deliver into global energy markets. Indeed, as shown in **Figure 1** for 2023, a majority of people in nearly every county in every state feel that way (with some states topping off above 80%).

Debates about the structure of a carbon tax or cap-and-trade regime continue, of course, but they are mostly about what to do with the revenue to ameliorate the regressive character of their incidence. Prasad [22], for example, reviews their content. Of particular note is the well-taken recognition of the danger in using carbon revenue to reduce other tax sources of government funds – the expectation that those proceeds would diminish over time if the policy is successful in reducing emissions.

What can be done? The United States is supposed to be a representative democracy, but the climate views of people who represent the Americans whose views are reflected in the Yale surveys are certainly not seeing positions being captured in either climate actions or even carbon policy design. Their contrarian representatives in the House and the Senate have been standing in the way for years, but that obstinance has been exaggerated by the new administration in the White House.

Scholars cannot give up pushing for what would work better in even the status quo because they face a political impediment born in the last century that still persists. They cannot stop pushing to price carbon nationally by putting climate change and climate action front of mind in every race being contested in every election, starting from the next one. And they can be buoyed by new evidence that policies

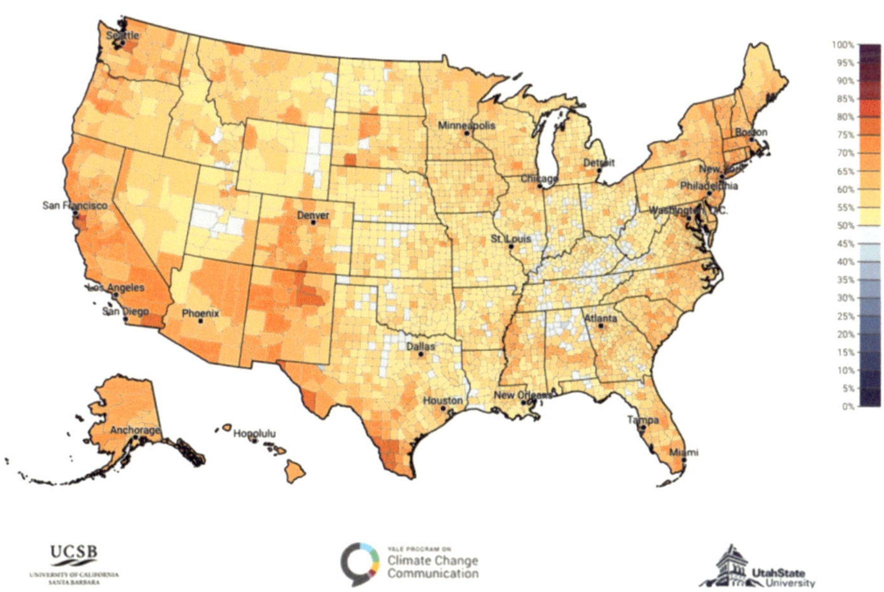

Figure 1.
County-level survey estimates of the proportion of residents who favor taxing fossil fuels entering the United States Economy. Blue – auburn boundaries differentiate counties on the 50% threshold and none of the blue counties are below 40%. Source: County map, Yale Program on Climate Change Communication [21], Prasad [22].

designed to promote emissions reductions work if somebody pays for doing otherwise [Stechemesser et al. [23]]. {SCIE} They must push at the ballots in every state where races for the President, Senators, and House Representatives can be found. They must continue to get the word out so that contrasts across candidates can be honestly assessed. They can celebrate every victory and mourn every loss for one day but then start all over again when they wake up the next day.

Author details

Gary Yohe
Wesleyan University, Connecticut, USA

*Address all correspondence to: gyohe@wesleyan.edu

References

[1] Weitzman ML. Prices vs. Quantities. The Review of Economic Studies. 1974;**41**(4):477-491. Available from: https://scholar.harvard.edu/weitzman/publications/prices-vs-quantities

[2] Yohe G. Towards a general comparison of Price controls and quantity controls under uncertainty. Review of Economic Studies. 1978;**45**:229-238. Available from: http://gyohe.faculty.wesleyan.edu/files/2018/05/6.pdf

[3] Yohe G. Single valued control of a cartel under uncertainty – A multifirm comparison prices and quantities. Bell Journal of Economics. 1977;**8**:97-111. Available from: http://gyohe.faculty.wesleyan.edu/files/2018/05/3.pdf

[4] National Research Council (NRC). Climate Stabilization Targets - Emissions, Concentrations, and Impacts over Decades to Millennia, National Academies of Sciences, Engineering, and Medicine. Washington, DC: The National Academies Press; 2011. DOI: 10.17226/12877

[5] Intergovernmental Panel on Climate Change (IPCC). AR4 Climate Change 2007, Synthesis Report. Cambridge UK: Cambridge University Press; 2007. Available from: https://www.ipcc.ch/report/ar4/syr/

[6] Hotelling H. The economics of exhaustible resources. Journal of Political Economy. 1931;**39**(2):137-175. Available from: https://www.jstor.org/stable/1822328

[7] Nordhaus W. To slow or not to slow: The economics of the greenhouse effect. The Economic Journal. 1991;**101**:920-937. Available from: https://www.jstor.org/stable/2233864

[8] National Academies of Science. Engineering, and Medicine (NASEM). America's Climate Choices – Final Brief. Washington: National Academies Press; 2011. Available from: https://nap.nationalacademies.org/resource/12781/ACC-final-brief.pdf

[9] Center for Climate and Energy Solutions (C2ES). Waxman-Markey Short Summary. 2009. Available from: https://www.c2es.org/document/waxman-markey-short-summary/

[10] European Commission. EU Missions Trading System (EU ETS). 2024. Available from: https://climate.ec.europa.eu/eu-action/eu-emissions-trading-system-eu-ets_en

[11] Yale Environment 360 (YE360). The Waxman-Markey Bill a Good Start or a Non-Starter. 2009. Available from: https://e360.yale.edu/features/the_waxman-markey_bill_a_good_start_or_a_non-starter

[12] United States Treasury (US Treasury). 2023 Inflation Reduction Act. Available from: https://home.treasury.gov/policy-issues/inflation-reduction-act

[13] United States Securities and Exchange Commission (SEC). SEC Adopts Rules to Enhance and Standardize Climate-Related Disclosures for Investors. 2024. Available from: https://www.sec.gov/news/press-release/2024-31

[14] Federal Reserve System. Dodd-Frank Stress Test Publications – 2023 Stress Test Scenarios. 2023. Available from: https://www.federalreserve.gov/publications/2023-Stress-Test-Scenarios.htm#:~:text=The%20Federal%20Reserve%20conducts%20stress%20

tests%20to%20ensure,and%20resulting%20capital%20levels%20under%20hypothetical%20recession%20scenarios

[15] CERES. 2023 Climate Scorecard. 2024. Available from: https://resources.ceres.org/ceres_scorecard/?utm_source=bing&utm_medium=paid&utm_campaign=accelerator_finreg-2023_lead-gen_keywords&utm_term=bing&msclkid=0145a9175cf91d3ab242cf25c019a601

[16] United States Environmental Protection Agency (EPA). Biden-Harris Administration Finalizes Suite of Standards to Reduce Pollution from Fossil Fuel-Fired Power Plants. 2024. Available from: https://www.epa.gov/newsreleases/biden-harris-administration-finalizes-suite-standards-reduce-pollution-fossil-fuel

[17] United States Department of Transportation (DOT). USDOT Announces New Vehicle Fuel Economy Standards for Model Year 2024-2026. 2022. Available from: https://www.nhtsa.gov/press-releases/usdot-announces-new-vehicle-fuel-economy-standards-model-year-2024-2026

[18] Carbon Credits. Carbon Prices Today. 2024. Available from: https://carboncredits.com/carbon-prices-today/

[19] Regional Greenhouse Gas Initiative (RGGI). Auctions. 2024. Available from: https://www.rggi.org/program-overview-and-design/elements

[20] Macrotrends. U.S. Carbon (CO_2) Emissions 1960-2024. 2024. Available from: https://www.macrotrends.net/global-metrics/countries/USA/united-states/carbon-co2-emissions

[21] Yale Program on Climate Change Communication (YPCCC). Yale Climate Opinion Maps 2023. 2023. Available from: https://climatecommunication.yale.edu/visualizations-data/ycom-us/

[22] Prasad M. Hidden benefits and dangers of carbon tax. In: PLOS Climate. 2022. Available from: https://journals.plos.org/climate/article?id=10.1371/journal.pclm.0000052

[23] Stechemesser A, Koch N, Mark E, Dilger E, Klösel P, Menicacci L, et al. Climate policies that achieved major emission reductions: Global evidence from two decades. Science. 2024;**385**:884-892. Available from: https://www.science.org/doi/10.1126/science.adl6547

Section 3

Resilience and Improving Our Abilities to Cope with Unavoidable Residual Harm

Chapter 5

Climate Risk, Climate Justice, and Why We Need Stronger Climate Governance

Paul Clements

Abstract

Scientists often assume that their analyses of risks will lead to rational action through the policy process. In reality, however, levels of risk normally depend on how information is channeled through political and institutional processes. Here, climate change presents great challenges. This chapter argues that climate change policy is best understood in terms of political economy and climate justice. The global nature of climate change confronts an anarchic system of nation-states in which agents who benefit from carbon pollution hold great power, while those at risk of residual harm tend to be politically weak. I frame the fundamental challenge of climate justice by showing revisions it calls for in *The Law of Peoples* proposed by John Rawls, the leading political theorist of the twentieth century. Whereas Rawls imagines domestic and international justice in separate spheres, climate justice requires an integrated framework with stronger international governance than Rawls is prepared to sanction. This framework is used to analyze the politics of mitigation, adaptation, and support for climate migrants in the United Nations Framework Convention on Climate Change, the limitations of present policies, and the global governance needed to reduce risks and promote climate justice.

Keywords: climate justice, UNFCCC, mitigation, adaptation, climate migrants, Rawls, original position, global governance, climate risk

1. Introduction

In 2022, Aramco, Saudi Arabia's state oil company, and ExxonMobil, the world's largest private oil company, earned record profits, respectively, of $161 billion and $56 billion [1, 2].[1] Also in 2022 floods submerged one-third of Pakistan, with thousands killed, over $30 billion in damages and economic losses, and more than eight million people pushed into poverty [3]. In 2023, after Somalia's rains failed for a fifth season, 3.8 million people—over a fifth of the population—were living as internally displaced persons, forced from their homes by drought and violence [4].

Back in 2007, the Intergovernmental Panel on Climate Change (IPCC) concluded that "[r]esponding to climate change involves an iterative risk management approach

[1] All financial figures in this chapter are given in US dollars.

that includes both mitigation and adaptation and takes account of climate change damages, co-benefits, sustainability, equity and attitudes toward risk [5]." Although this finding has framed much of the subsequent negotiations on climate change policy under the United Nations Framework Convention on Climate Change (UNFCCC), to think that this risk management process is determined by rational, science-based assessment of costs and benefits would be a mistake. Rather, the political power of beneficiaries like Aramco and ExxonMobil and the weakness of victims like Pakistani farmers and nomadic Somali pastoralists are important factors in systems for planning mitigation and adaptation.

From a social justice perspective, climate change looms largely as a set of negative international and intergenerational externalities, as beneficiaries of greenhouse gas (GHG) emissions in richer countries impose costs on victims in poorer countries and, in all probability, on future generations. Oil company executives are likely to perceive different risks from climate migrants and to have greater powers to influence these risks. Since carbon pollution is global, the most significant planning for managing it and its effects—planning for mitigation and adaptation—takes place at the global level, formally through the UNFCCC. The international system of nation-states, however, is anarchic, with each state promoting its own national interests. The determination of national interests, in turn, depends on power structures within each country. While major carbon polluters often have significant influence on the articulation of national interests, victims of climate change tend to be concentrated among politically weaker sections in politically weaker countries. This has led to mitigation and adaptation policies that fail to hold polluters accountable for the harm they impose on others and that also fail to adequately protect the interests of current victims and increase risks for vulnerable populations in the future.

Unfortunately, there is no way to empower climate change victims in future generations, and it would be hard to empower current victims in low-income countries. In this context, the fundamental cause of weak mitigation and adaptation needs to be traced back to the anarchic system of nation-states, and solutions need to be found to strengthen global governance. To enforce fair mitigation policies and to protect the rights, for example, of climate migrants, the UNFCCC needs to build institutions that can enforce fair limits on carbon polluters, secure adequate resources from recalcitrant countries, and establish much more significant protections and supports, especially in more vulnerable countries. While this agenda presents great political challenges, it represents the most viable pathway to reducing future risks and achieving a degree of climate justice.

2. A Rawlsian approach to climate justice

To appreciate the challenges posed by climate justice it is important to recognize that domestic and international justice normally operate in separate spheres. The degree of separation can be seen in how John Rawls, often considered the leading political theorist of the twentieth century, frames the problem of social justice for each sphere.

In his groundbreaking book, *A Theory of Justice* (1971), Rawls asks what principles free and equal individual persons in a constitutional democracy would adopt to regulate costs and benefits arising from their social cooperation. To address this question, he argues that we should imagine that we do not know our position in society—whether we are rich or poor, male or female, black or white, and so on. He argues that from this "original position" we would want to establish conditions in which all citizens have access to basic resources needed to pursue a life they choose, and this reasoning leads to his well-known principles of justice:

- *First*: Each person is to have an equal right to the most extensive scheme of equal basic liberties compatible with a similar scheme of liberties for others.

- *Second*: Social and economic inequalities are to be arranged so that they are both (a) reasonably expected to be to everyone's advantage, and (b) attached to positions and offices open to all [6][2].

In Rawls' conception of a just society, while the courts protect freedoms of speech, belief, and political participation, the legislature provides for education, health care, and a basic income. To defend equal opportunity and the fair value of liberties, the legislature also establishes provisions to avoid excessive inequalities, such as through progressive income and inheritance taxes [6].

While *A Theory of Justice* proposes principles of justice for individual countries, in *The Law of Peoples* (1999) Rawls follows a similar approach to propose principles for international relations. Now to identify principles for free and equal *countries*, to "protect their political independence and their free culture with its civil liberties, [and] to guarantee their security, territory, and the well-being of their citizens [7]," Rawls imagines a second original position. In this case, rather than not knowing our position in society, we are to imagine that principles are chosen as if we did not know the relative position of each country: the size of its territory or population, its economic or military strength, or its natural resources.

In this second original position, Rawls argues that eight familiar and traditional principles of justice among free peoples would be affirmed. Most relevant to climate justice are:

1. People are free and independent, and their freedom and independence are to be respected by other people.

2. People are to observe treaties and undertakings.

4. People are to observe a duty of nonintervention.

8. People have a duty to assist other people living under unfavorable conditions that prevent them from having a just or decent political and social regime [7].

These principles that Rawls proposes for domestic and international justice and his approach to developing them represent the most prominent contribution to the philosophical discourse on justice in the recent past. Also, his framework, with one system for domestic and another for international justice, reflects the basic structure of the world order. Countries or states are the primary units for international justice as individual persons are for domestic justice. In international relations, how each country manages its commitments internally, distributing costs and benefits among its citizens, is normally not an issue for other governments.

[2] This is his initial statement of the principles. They are further developed through the course of the book.

We can note that while supporting everyone's participation in the system, Rawls' principles for both domestic and international justice give particular attention to those who are less advantaged. In the domestic case, for social and economic inequalities to be to everyone's advantage requires more advantaged members of society to promote their own interests in ways that also help others, and in the international case, the duty of assistance requires stronger states to support weaker states. In both systems we anticipate significant inequalities, but to support the participation of all, justice requires particular attention to those who are more vulnerable.

When it comes to climate change, however, the well-established framework that separates international from domestic justice is deeply inadequate. It is individuals who directly experience costs and benefits from GHG emissions, but national governments negotiate the central framework for managing these costs and benefits. Once we take account of power relations, it is clear that the interests of victims are likely to be underrepresented twice, within each country and in negotiations between countries. Heavy polluters, however, are likely to avoid responsibility twice, among citizens and again among countries.

Given the distribution of costs and benefits from GHG emissions, to identify principles for climate justice we need to construct a third original position, once again populated with individual persons as in the domestic case, but now imagining that we do not know what country we live in or our social position. Wealthier persons and countries tend to have achieved their wealth partly from activities that produce higher cumulative emissions, while harms from these emissions tend to fall more heavily on persons and countries with lower emissions. What principles for climate change mitigation and adaptation would be fair to Pakistani farmers, to nomadic Somali pastoralists, to oil company executives, and to future generations?

The "veil of ignorance" is the image Rawls uses to represent constraints on information in the original position leading to the selection of principles that are fair. Behind the veil, besides not knowing our place in society for domestic justice or our country for international justice, the "time of entry" is also hidden. For our third original position, besides now not knowing our place in the world, this means selecting principles as if we also did not know the date, from now through the indefinite future. Principles to be applied today should be acceptable to people at any future time, such that for mitigation, for example, the principles should describe an emissions trajectory that is justifiable at each time. This is complicated by uncertainty in levels of risk associated with different emissions trajectories, requiring, as the IPCC notes, "iterative risk management," based both on effects revealed over time and on evolving projections as knowledge improves. The UNFCCC and other authorities generally rely on IPCC projections for risk estimates, but these tend to have wide confidence intervals, and the IPCC's most recent assessment report (AR6 in 2021) projects greater risks than earlier reports [8]. A significant minority of climate scientists, notably James Hansen [9], argue that true risks are mostly greater than IPCC estimates, while many policymakers underestimate or fail to grasp risks due to climate science denial and/or to conceptual and analytic complexities of risks and policy responses [10].

In the original position for climate justice, we consider the structure of costs and benefits from GHG emissions starting from the present situation. Historic emissions have increased global atmospheric concentrations of carbon dioxide from 280 to 419 parts per million from pre-industrial times to 2023, and annual emissions continue to rise [11]. We need to identify a global carbon budget for additional GHG emissions, a just pathway for reducing emissions to net-zero, and plans for adaptation and for addressing residual harms. Failure to reduce GHG emissions increases the need for adaptation and failed adaptation increases residual harm, so institutions in all three areas must be considered as a unified system. Following the UNFCCC, we focus on two sets of countries—developed or industrialized and developing (including most small island states)—but we also attend to gains and risks among individuals. As with domestic and international justice, an original position for climate justice highlights concern for the more vulnerable. As a first principle, not knowing our position in the world, it is clear that we would want to protect persons whose basic needs climate change threatens. Everyone is at risk from climate change, but victims whose life prospects are severely harmed or at greater risk particularly deserve institutional support.

3. The UNFCCC is inadequate for just mitigation

The current international agreement for managing climate change, the Paris Agreement, aims to hold "the increase in the global average temperature to well below 2°C above pre-industrial levels and [to pursue] efforts to limit the temperature increase to 1.5°C above pre-industrial levels [12]." According to IPCC projections, this implies a carbon budget from 2024 of 353 Gt for 1.5°C and 1003 Gt for 2°C, while annual emissions were 37.5 Gt in 2023 [13, 14]. Given harm from climate change to innocent victims, in our third original position I will argue that we would want to endorse this goal. There is considerable evidence, however, that the 1.5°C target is no longer feasible in any plausible scenario. While it is technically feasible to reach the 2°C target, its institutional feasibility remains uncertain. Institutional feasibility depends, however, mainly on levels of effort from industrialized countries to reduce their own emissions and to support developing countries in reducing theirs. 2°C probably would be feasible if industrialized countries were held responsible for adaptation costs and residual harms that they impose on others. The central political inadequacy of the UNFCCC is found in its failure to impose this responsibility.

Political foundations of weak mitigation have been evident since talks leading up to the establishment of the UNFCCC in 1992. In the Intergovernmental Negotiating Committee (INC) that prepared the text for the UNFCCC, small island states, particularly vulnerable to sea level rise, tried to include a principle of responsibility based on GHG emissions, but this was adamantly opposed by industrialized countries, particularly by the United States.

> *Throughout the INC, industrialized countries maintained their strong opposition to accommodating any notion of responsibility, liability, or compensation, and remained unswayed both by moral and formalistic legal arguments. With the primary goal of securing the participation of the US in the future climate change regime as the then largest GHG emitter, developing countries were slowly forced to soften their stance on responsibility [15].*

The phrase "common but differentiated responsibilities" (CBDR) finally adopted in the 1992 convention avoids assigning legal responsibility. In clarifying the meaning of CBDR, the convention states: "Accordingly, the developed country Parties should take the lead in combatting climate change and the adverse effects thereof [16]," and that this includes financial and technical support to developing countries, but only as statements of principle.

Since 1992, therefore, although industrialized countries grown rich with higher GHG emissions are expected to take the lead in mitigation and adaptation, there have been no enforced limits on further emissions or mandated support for victims of adverse effects. It was not until 2015 when the 21st meeting of the Conference of Parties to the UNFCCC (COP 21) reached the Paris Agreement, that targets for limiting global warming were finally quantified, and this agreement continues to avoid legal requirements. Instead, it allows each participating country to set its own "nationally determined contribution" to reducing GHG emissions. The agreement recognizes the need to support developing country Parties [12] but with no consequences for limited action in mitigation or adaptation. An addendum clarifying the terms of the Paris Agreement explicitly states that "the Agreement does not involve or provide a basis for any liability or compensation [17]."

The original 1992 convention actually does state an aim for the developed countries to return their GHG emissions "individually or jointly to their 1990 levels" by the year 2000 [16]. A plan to implement reductions in developed country emissions, however, was not completed until 1997 when COP 3 proposed the Kyoto Protocol, and in the interim American oil companies became aware of the threat such policies would pose to their profits. Once it became clear from negotiations leading to the Kyoto Protocol that it would require developed countries to reduce their GHG emissions, ExxonMobil and other oil companies spent millions of dollars building a coalition to lobby the US Senate against it [18]. Although then-US President Clinton signed the Kyoto Protocol, and despite the US being a Party to the 1992 UNFCCC, the US Senate refused to ratify it [19]. The absence of the leading superpower took the air out of the Kyoto Protocol process, and global GHG emissions continued to increase almost every year not only from 1992 to 2000 but through to 2021 when the Covid-19 pandemic caused a temporary reduction [20].

It is impossible to untangle the influence of oil companies, other high-carbon polluters, and governments of high-polluting countries on these governments' rejection of responsibility for their national pollution. While the US and major oil exporting countries like Saudi Arabia are routinely cited as opposing strong limits on GHG emissions and legal responsibility for their effects [21], interests of individuals, firms, and governments in financial and economic gains and in political power overlap in contributing to opposition to legal responsibility.

Countries with strong political commitments to climate action such as Denmark, Scotland, and Sweden demonstrate that it has been possible to implement Paris Agreement-consistent energy policies without compromising national living standards [22], and the International Energy Agency has published a roadmap showing how it is technically feasible for the international community to achieve net-zero carbon emissions by 2050 without significantly harming economic growth [23]. With good management, heavy carbon polluters could be required to reduce emissions in line with the IPCC's carbon budget for the 2°C target without suffering inordinate harm, but great suffering from the effects of climate change cannot be avoided even with the greatest feasible efforts at prevention. The 1992 Convention and the Paris Agreement already articulate broad principles of rights and responsibilities among developed and developing countries that would largely be supported in the original position. Where these principles fall short is in failing to specify detailed responsibilities, failing to sanction noncompliance, and failing to address the needs of individual victims. Demands in these areas, in turn, require stronger institutions than the UNFCCC has so far established. It probably is not feasible to hold developed countries legally liable for their full share of responsibility for adaptation costs and residual harms based on levels of carbon pollution (even if this responsibility could be specified), but climate justice depends on connecting polluters' responsibilities to victims' levels of risk. A central task from the original position for climate justice is to sketch this connection.

Victimization from climate change involves harms already experienced, harms "in the pipeline" due to inevitable increases in atmospheric levels of GHGs, and risk profiles for future climate scenarios. Although unprecedented heat waves, droughts, floods, and other extreme weather events have become the stuff of our daily news, as other chapters in this volume attest, the state of knowledge does not permit precise aggregation of harms due to climate change. For example, Ballester et al. estimate 61,672 heat-related deaths during Europe's hottest season on record in 2022 [24], but while all of India's warmest years on record have occurred in the past decade, de Bont et al. attribute only about 1116 Indian deaths a year to heatwaves [25], and public health experts note that heat-related deaths are systematically undercounted in India [26]. A brief review of the evidence demonstrates, however, along with examples discussed above, that victimization from climate change is already substantial and will become more so.

Bangladesh suffers from climate-induced displacement due to coastal hazards from storms and sea level rise and due to river erosion and flooding. Based on household survey data from the Bangladesh Bureau of Statistics, Ahmed estimates that there were about two million permanent climate migrants within Bangladesh from 2001 to 2011 [27], and families forced from their homes by storm surges or flooded rivers typically face dire conditions as they resettle in urban slums [28, 29]. At the global level, Li, Yuan, and Copp find that with 1°C global warming, the number of people exposed to life-threatening combinations of heat and humidity at least one day per year has risen from 97 million to 279 million, and it projected to rise to 508 million at 1.5°C [30]. The annual number of recorded internally displaced persons (IDPs) worldwide never exceeded 10 million from 2000 to 2005, but since then it has risen almost every year to 57 million in 2022 [31]. These include internal displacements (excluding people who cross national borders) due to both conflict and violence and to weather disasters, but weather disasters typically makeup at least half the total and sometimes contribute to conflict and violence. Bilal and Känzig find worldwide macroeconomic damages from climate change six times larger than most previous estimates, indicating "that world GDP per capita would be 37% higher today had no warming occurred between 1960 and 2019 instead of the 0.75°C observed increase in global mean temperature [32]." While earlier economic studies are based on variations in local temperatures, they find increases in global temperature to cause much greater economic harm. The strongest negative effects are found "in relatively hot regions such as Southeast Asia and Sub-Saharan Africa," while Central and East Asia, with many colder countries, experienced economic gains [32].

Although methodological questions can be raised about quantitative estimates of climate change impacts from each of these examples, together they demonstrate a burden of past victimization that cannot be justified, and risks will inevitably increase as global temperatures continue to rise. From the perspective of the original position, with developed countries and China having contributed over 70 percent of historic carbon emissions [33], global GHG emissions on their current trajectory are clearly unjust. Future climate change scenarios reinforce the urgency of achieving net-zero emissions and subsequently removing GHGs from Earth's atmosphere. While die-offs of the world's coral reefs [34] and the Amazon rainforest [35], a slow-down of the Gulf Stream [36], melting glaciers [36], and reductions in biodiversity [37] are among many potentially catastrophic impacts, perhaps the most iconic long-term risk is found in rising sea levels.

The IPCC's AR6 finds that "[c]onsidering only processes for which projections can be made with at least *medium confidence*," gross mean sea level is likely to rise by between 0.28 and 1.01 meters by 2100 relative to the period 1995–2014 and by up to 16 meters by 2300 in the most unfavorable scenario. The report notes, however, that sea level rise involves deep uncertainties, particularly in the speed at which Antarctic ice sheets may collapse, and it offers a storyline in which sea level rises by 2.3 meters by 2100 [36]. A meter of sea level rise would devastate cities across the world home to many millions of people, particularly in Asia [38]. The Mekong Delta, Vietnam's breadbasket and home to 21 million people, has an average altitude of less than a meter above sea level and would be largely submerged [39]. Hansen et al. argue, however, based in part on sea level rise of 6 to 9 meters above current levels in the most recent interglacial period when Earth was less than 1°C warmer than today, that sea level rise could reach several meters over a timescale of 50–150 years [40]. Physical processes leading to sea level rise simply are not well understood, and, as in other areas, it appears that uncertainties in IPCC projections fall mainly on the dangerous side. Several meters of sea level rise in this century would be economically catastrophic for many countries and the forced migrations it would cause would create unprecedented social tensions. Responsibilities to people in future centuries also weigh heavily, as we do not know at what point many meters of sea level rise become unavoidable.

Based on past and ongoing victimization and future risks, the Paris Agreement target to keep global warming well below 2°C would clearly be endorsed from an original position for climate justice, but the system of nationally determined contributions (NDCs) is unlikely to achieve this target and the distribution of GHG emissions is unlikely to be just. Emissions reductions are a global public good, but with each country setting its own national target and no sanctions for missing targets, levels of commitment depend on each country's internal politics or political pressures outside the UNFCCC framework. The UN Environment Program estimates that globally, current policies lead to about 3°C global warming, falling to 2.5°C if all countries meet both conditional and unconditional commitments in their NDCs [41]. Conditional commitments, mostly by developing countries, depend on financial and/or technical support from developed countries. Of 168 NDCs submitted by 2016, mitigation targets for 110 were wholly or partially conditional [42].

Developed countries are responsible for reducing their own emissions and for helping developing countries to reduce theirs. The potential fragility of developed countries' mitigation commitments is demonstrated by US President Trump's 2017 move to withdraw the US from the Paris Agreement [43] and his canceling mitigation policies from the prior Obama administration [44]. Many European countries have seen political backlashes against policies to transition to renewable energy, and when Russia cut natural gas supplies to Europe (hoping to weaken European support for Ukraine after Russia's invasion) many European countries "return[ed] to their coal plants with a vengeance," reversing prior reductions in GHG emissions [45].

With the Paris Agreement, developed countries reaffirmed their commitment to provide $100 billion in annual support for mitigation and adaptation by developing countries from 2020, with funding to be balanced between mitigation and adaptation. Pauw et al. estimate, however, that it would cost $2.8 trillion to fund all conditional mitigation commitments in NDCs from 2021 to 2030 [44], compared to $0.5 trillion if $50 billion a year were devoted to NDCs' mitigation in line with Paris commitments. In the event, developed countries' reported climate financial commitments did not reach $100 billion until 2022, and Oxfam finds that due to dishonest and misleading accounting, "[t]he true value of climate finance is a third of what developed countries report [46]." While the quality and comparability of NDC estimates do not support reliable aggregation, funding commitments for developing countries' mitigation are clearly inadequate.

From the original position for climate justice developing countries' economic development and poverty reduction goals retain priority with respect to mitigation goals, and this is also the view taken by the 1992 convention and the Paris Agreement. The International Energy Agency's roadmap for net-zero emissions by 2050 has global electricity demand more than doubling by 2050, as electricity substitutes for carbon-based fuels but also as many households in developing countries gain access to electricity for the first time and as industry expands [23]. Besides being underfunded, however, institutions for implementing developed countries' support for developing countries' mitigation are critically insufficient and often inefficient.

For the most part, developed countries' support for climate action in developing countries is channeled through dozens of pre-established bilateral and multilateral development agencies, such as the US Agency for International Development, the European Bank for Reconstruction and Development, and the Global Environment Facility housed at the World Bank. This structure precludes rational management of the enormous expansion of institutional capacity needed to build developing country capacities to implement mitigation goals, and the agencies' political and organizational interests often conflict with development goals. Funds are usually committed in response to project proposals, but (as we discuss below) the most vulnerable countries typically have weaker infrastructure for preparing proposals and for project implementation.

An obvious project metric would seem to be the cost per ton of carbon dioxide emissions avoided (often in the context of increasing electricity use), but this metric is often not found in project proposals. For example, neither plans for the Global Environment Facility's project for Promoting Low-carbon Electric Bus Transport in Mauritius [47] nor for the African Development Fund's Mini Grid and Solar PV Net Metering project in Ghana [48] presents the cost per ton of carbon dioxide emissions avoided. The project in Mauritius aims to replace 30 diesel busses with electric busses at a cost of $21 million. It expects to avoid 10,950 metric tons of carbon dioxide emissions as direct benefits and another 18,250 metric tons in indirect benefits as additional electric busses are added in the future. The project in Ghana aims to install 12,000 solar rooftop PV units and to build 35 mini-electric grids at a cost of $85 million, with 718,590 tons of carbon dioxide emissions avoided per year compared to the alternative powered by oil and natural gas. With a standard estimate of 25 years of benefits for solar rooftop units, the project would avoid 17,964,750 tons of emissions over its lifetime. Without discounting, the cost per ton of emissions avoided by the electric bus project in Mauritius is $720, compared to about $5 for the solar rooftop PV and mini-grid project in Ghana. These calculations indicate that increasing solar units in countries like Ghana is likely to be far more cost-effective in mitigating carbon emissions than switching from diesel to electric busses in countries like Mauritius, but the donor agencies are not highlighting data that would support such conclusions.

At this writing, global temperatures have exceeded 1.5°C above pre-industrial levels over the last 12-month period [49]. Since Paris targets are based on long-term averages this does not imply that the 1.5°C target has been breached, but it supports the view that IPCC carbon budgets are optimistic. If we accept the IPCC's 1003 Gt carbon budget for the 2°C target as a working estimate, from an original position for climate justice we would want to divide this budget among the world's countries on some fair basis and apply sanctions to countries that exceed their national budget. Developed countries would also be required to provide adequate funds to support developing countries' transitions to renewable energy, as well as funding adaptation and efforts to address residual harms as discussed below. Developing countries' carbon budgets would come with the proviso that they receive adequate financial and technical support.

While the original position for climate justice focuses on the conditions of individual persons, for mitigation it is reasonable to address the obligations of national governments. Given political realities, national governments are unlikely to impose mitigation costs fairly, but the UNFCCC could not be expected to supervise national tax codes or infrastructure development plans. The domestic justice of national mitigation plans is grounded in the wider social justice (or injustice) of national institutions.

In order to implement sanctions for noncompliance and support developing countries' mitigation, the UNFCCC and its supporting organizations need greater legal powers and institutional capacity than they currently possess. As they also need greater capacity to support adaptation and to address residual harms, these matters are addressed together in our conclusion. The greatest risks from climate change come from failing to limit carbon emissions with sufficient and secure national commitments.

4. The UNFCCC is inadequate for just adaptation

Whereas mitigation aims to reduce the extent of climate change, adaptation aims to limit its risks to persons, whether by preventing the effects of climate change from causing harm or by reducing levels of harm that nevertheless occur. Since mitigation involves incurring national costs for a global public good, it often requires external motivation such as sanctions. Adaptation, by contrast, aims to reduce risks to actors within a national territory. In principle, national actors should be willing to undertake adaptation efforts up to the point at which their costs exceed the savings they yield.

The extent of adaptation that is needed depends, of course, on global levels of atmospheric GHGs that come mostly from industrialized countries. Besides industrialized countries' being unwilling to fund all the adaptation they have necessitated, I will argue that channels of support from developed to developing countries also do not allocate adaptation resources fairly or efficiently. Also, whether in developed or developing countries, adaptation often requires long-term investments that cross jurisdictions and involve great uncertainties. Climate change undermines land and property values, but adaptation is inevitably uneven in whose assets it protects or restores. Political processes underlying the allocation of adaptation resources are likely to favor established powers, particularly in more vulnerable countries.

The weak focus on more vulnerable persons in support of adaptation, and, as we discuss below, the weak support for addressing residual harms from climate change, offer the strongest evidence of the injustice of the anarchic, state-based system that is the UNFCCC. These weaknesses also demonstrate the inadequacy of the state-based focus of Rawls' *Law of Peoples* for climate justice. A conception of a just law of peoples must also idealize the roles of governments in representing their peoples' interests. Since international relations are mainly worked out by national governments, an ideal conception of international relations such as found in a conception of a just law of peoples necessarily imagines as a counterpoint domestic justice among people who support this law. UNFCCC agreements similarly assume implicitly that each government will implement its domestic obligations, if not with perfect justice, at least without excessive injustice.

Among the principles in Rawls' *Law of Peoples*, there is "a duty of assistance to other peoples living under unfavorable conditions that prevent their having a just or decent political regime" (Principle #8). Contrary to some analyses by scholars of Rawls (e.g., [50]), this is not the source of the obligation to support individual persons victimized by climate change. Since their victimization is largely caused by unjust levels of GHG emissions from developed countries, the obligation arises rather from these countries' violation of Principle #4, that "[p]eoples are to observe a duty of non-intervention." In light of normal political processes, however, governments of developing countries cannot generally be relied upon adequately to represent the interests of persons who face greater risks among their citizens. In this context climate justice needs to be worked out not from the perspective of peoples or states as *The Law of Peoples* has it, but from the perspective of individual persons, as in our third original position. Here we see that the obligation to protect the interests of victimized persons reverts to the international body governing responses to climate change, the UNFCCC.

It is widely recognized, however, that the UNFCCC has failed to mobilize adequate resources for adaptation. The UN Environment Program, in its Adaptation Gap Report 2023 entitled *Underfinanced. Underprepared. Inadequate investment and planning on climate adaptation leaves the world exposed*, finds that:

> *the adaptation finance gap is widening and now stands at between US$194 billion and US$366 billion per year. Adaptation finance needs are 10–18 times as great as current international public adaptation finance flows – at least 50 percent higher than previously estimated [51].*

Similarly, the Global Commission on Adaptation led by Ban Ki-moon, 8th Secretary-General of the United Nations, Bill Gates, then co-chair of the Bill and Melinda Gates Foundation, and Kristalina Georgieva, then CEO of World Bank, in its 2019 flagship report entitled *Adapt now: A global call for leadership on climate resilience*, finds that although "investing $1.8 trillion globally in five areas from 2020 to 2030 could generate $7.1 trillion in total net benefits," in 2015–2016, "global public financing for adaptation was [only] about $22 billion per year [52]." The Climate Policy Institute, in its comprehensive assessment of global climate finance in 2021/2022 including both public and private sources, finds that from the total average annual climate finance of $1.27 trillion, only $63 billion, about five percent, was devoted to adaptation, compared to $1150 billion for mitigation and $51 billion for "overlap" investments with dual-benefits [53]. The IPCC's AR6 finds that despite progress since AR5,

> *Most observed adaptation is fragmented, small in scale, incremental, sector-specific, designed to respond to current impacts or near-term risks, and focused more on planning rather than implementation (high confidence). Observed adaptation is unequally distributed across regions (high confidence), and gaps are partially driven by widening disparities between the estimated costs of adaptation and documented finance allocated to adaptation (high confidence). The largest adaptation gaps exist among lower-income population groups (high confidence). At current rates of adaptation planning and implementation, the adaptation gap will continue to grow (high confidence) [54].*

While climate change requires adaptation across the spectrum of human activities and AR6 discusses adaptation "options" in 23 areas, the Global Commission on Adaptation's breakdown of adaptation programs into six areas offers a convenient summary. These include (1) food security and livelihoods of small-scale producers, (2) natural environment, (3) water, (4) cities and urban areas, (5) infrastructure, and (6) disaster risk management [54]. For example, increasing drought and heat threaten farmers around the world, and the possibility of multiple simultaneous production crises threatens global food security. Protecting forests and mangroves provides important environmental services while also sequestering carbon. Mismanagement of agricultural and urban water systems has long been widespread, offering significant opportunities for reform. Defenses can be built to protect cities from sea level rise, stronger storms, and heat waves, and roads, airports, seaports, and electric grids can be upgraded. Precautions can be taken everywhere to prevent hazards from becoming disasters.

In 2001 the UNFCCC started to encourage and support the least developed countries to develop National Adaptation Programs of Action (NAPAs) and in 2010 it launched a National Adaptation Plan (NAP) process with developed and developing countries [55]. It supports funding for developing countries' adaptation through a Least Developed Countries Fund (2002), a Special Climate Change Fund (2004), an Adaptation Fund (2008), and a Green Climate Fund (2015) that also funds mitigation projects [56] (year operationalized in parentheses). All these funds are managed by the World Bank, the first two through its Global Environment Facility, and the World Bank also independently funds many climates change-related projects. Despite the substantial role of the World Bank and other multilateral agencies in adaptation finance, the majority of official adaptation finance is channeled through bilateral development agencies [53]. Beyond official finance, many people respond autonomously to changing climatic conditions (for example, see Ref. [57]). It is reasonable to expect that considerable adaptation is undertaken independently by households, firms, and sub-national government units and that most of this is not reflected in statistics on adaptation finance.

The overwhelming majority of official adaptation finance is channeled through projects managed largely in accordance with long-established organizational practices of the official international development community. While donor agencies have incentives to "move money" and typically aim to expand their budgets (the classic analysis is [58]), governments and a host of other organizations based in developing countries often depend on accessing donor funds. Given the systematic properties of the energy sector, mitigation generally lends itself to rational planning. By contrast, it is possible to imagine almost any development project as contributing somehow to adaptation. Proposals requesting adaptation resources may often be opportunistic, and it is often difficult to estimate a proposed project's contribution to reducing risks. Not only are adaptation funds greatly insufficient; but bureaucratic systems for managing these funds are also generally too weak to allocate resources efficiently with respect to risks.

Since most adaptation funds are allocated in response to project proposals, governments and other entities that are better at writing proposals and with greater project implementation capacities tend to get more funds. As Garshagen and Doshi find in their analysis of the Green Climate Fund:

> *the proposal process results in the fact that many countries with the highest climate vulnerability but weak government institutions and fragile state bureaucracies*

> *have missed out and not been able to access project funding, mostly least developed countries (LDCs) in Africa and conflict-ridden countries. Further, most countries have not yet been able to access project funds independently through their national entities, limiting direct access and country ownership – the strengthening of which is a major goal of the fund [56].*

In his analysis of funding for over 100,000 climate action projects in 133 countries from 2000 to 2018, Islam finds that "countries within the Sub-Saharan Africa and South Asia regions, despite their higher climatic vulnerabilities, were likely to receive significantly less adaptation and overlap fundings." Also, the countries most at risk in his analysis, including DR Congo, Solomon Islands, Mali, Liberia, Sudan, Chad, Guinea-Bissau, Micronesia, Niger, and Somalia, tended to receive less adaptation and overlap funds than moderately vulnerable countries. Islam finds that more vulnerable countries often failed to present their funding proposals convincingly to donors and that their "ability to successfully implement climate-related interventions" was often questionable [59].

Although the UNFCCC and donor agencies have carried out many programs to strengthen adaptation planning and implementation capabilities in developing countries, as the quote above from the IPCC's AR6 suggests, they are far from sufficient for effective risk management. Given weak political representation, uneven bureaucratic capacities, and widespread corruption often found in low-income countries, challenges in programming adaptation resources should not be underestimated. In a comprehensive review of the literature on adaptation projects including 34 empirical studies of adaptation implementation, Eriksen et al. find that "[e]vidence for elite capture and manipulation exists from around the world [60]." Many projects, they find, "inadvertently reinforce, redistribute, or create new sources of vulnerability" through four mechanisms:

- Shallow understanding of the vulnerability context
- Inequitable stakeholder participation in both design and implementation
- A retrofitting of adaptation into existing development agendas, and
- A lack of critical engagement with how 'adaptation success' is defined

In this context, they argue that "learning processes within organizations and with marginalized populations must be placed at the centre of adaptation objectives [60]." Pisor et al. argue that standard, top-down donor approaches to project design often led to unsustainable outcomes and that both the equity and the efficiency of climate change adaptation depend on involving frontline communities themselves in developing adaptation strategies [61]. While most of Eriksen et al.'s and Pisor et al.'s concerns about donor agency practices are not new, their relevance is magnified by the relatively great management challenges inherent in the adaptation "sector".

Inadequate adaptation funding and inefficient programming lead to greater residual harm from climate change. While our focus has been on adaptation in developing countries, developed countries also face adaptation challenges. On the whole, however, particularly in light of their greater vulnerability, the depth and breadth of residual harms to persons is likely to be far greater in developing countries. In the original position for climate justice, we find that the UNFCCC must require developed countries to provide much greater funding for adaptation and that systems for supporting adaptation need to be rethought and redesigned.

5. The UNFCCC fails to support justice for victims of residual harms from climate change

The inadequacy for climate justice of both Rawls' *Law of Peoples* and the UNFCCC is particularly due to their weak representation of the interests of persons. In both cases, representatives of governments or of peoples are the main negotiating agents, hypothetically in *The Law of Peoples* and in practice in the UNFCCC. The UNFCCC's actual failures in defending the interests of persons illustrate the routine inadequacy of vulnerable states in representing the interests of their more vulnerable citizens.

In the original position for climate justice, I have argued that a first principle is to defend basic needs. Not knowing our place in the world or in time, we would prioritize avoiding conditions that place our fundamental well-being at great risk. While this first requires mitigation, institutions for adaptation and for addressing residual harms would be viewed together in terms of risk profiles and institutions' collective efficiency in reducing these risks. As resources are insufficient to address all harms, basic needs in principle have priority. Although basic needs need to be defined and there is no simple metric for comparing different harms from climate change, this principle has wide and intuitive applications. For example, the poor should generally be defended before the nonpoor, and the health of pregnant and lactating women and young children should often be prioritized. Recognizing the significance of cultural resources, small island states vulnerable to sea level rise should also often receive priority.

Beyond direct harm to persons, residual harms from climate change (occurring despite adaptation efforts) include economic harm to states, such as Pakistan's economic losses from the 2022 floods. From the original position, these harms need to be assessed in terms of how they undermine state capacities to address residents' basic needs. There is no simple formula for comparing these harms to direct harms to persons, but assessing the relative priority of economic harms is among the judgments climate justice requires.

Besides the loss of life, the most substantial risk to persons from climate change arguably arises from forced displacements. While climate change drives many deeper into hunger, poverty, and ill health, and risks from heat waves, water shortages, and uncertainty and anxiety are clearly significant, the loss of home, livelihood, and community commonly caused by forced displacement, and the harsh conditions that often follow it, are particularly severe harms. For example, comparing climate-induced and non-climate-induced migrants almost a decade after they arrived in Dhaka's Korail slum, in the capital of Bangladesh, Adri and Simon find that climate-induced migrants were living on $0.60 per person per day on average compared to $1.80 for non-climate-induced migrants. Climate migrants faced greater hunger and unemployment, had far fewer assets, tended to have more menial jobs, and all wished to return to their home villages [29].

Climate migrants worldwide face predictable deficits in basic needs, and programmatic challenges in addressing these needs tend to be less severe than in adaptation generally. It is important to recognize in this context that restoring climate migrants' capacities to secure decent livelihoods is critical not only to their own well-being but also to their ongoing participation in and contributions to the wider society. It is not their choice to leave their homes, and many arrive destitute with livelihood skills unsuited to their new environments. No doubt a fortunate few can launch new careers unaided, but many are left to unskilled employment and mired in poverty. As human potential is a collective as well as private good, restoring this potential not only corrects an injustice but also serves the community interest.

Despite limitations in data on climate migration, we know that there are tens of millions of climate-displaced persons worldwide and that displacements are sure to increase in the coming years. Nevertheless, the UNFCCC has largely ignored their needs. Compared to $100 billion per year for mitigation and adaptation, the UNFCCC's Task Force on Displacement established with the Paris Agreement has only 14 members and no program budget [62]. In principle, the needs of climate migrants could be addressed under adaptation, but in practice, adaptation has generally been understood as addressing more direct effects of climate change. None of the 513 priority projects from National Adaptation Programs of Action [63], 229 projects supported by the UNFCCC's Adaptation Fund [64], or 2459 climate change projects in the Global Environment Facility's projects database [65] has a title indicating support for climate migrants or persons displaced by climate change. (The Adaptation Fund supports a project titled "Increasing the resilience of both displaced persons and host communities to climate change-related water challenges in Jordan and Lebanon," but it supports people displaced by conflict). In 2023 the UNFCCC launched a Loss and Damage Fund that aims to "provide support for responding to economic and non-economic loss and damage associated with the adverse effects of climate change." While its terms of reference include addressing challenges from displacement, relocation, and migration, they also include challenges from climate-related emergencies, sea level rise, insufficient climate information and data, and climate-resilient reconstruction and recovery [66]. Also, with only $661 million in commitments as of March 2024 [67], the institutional strength of this fund remains uncertain.

In the absence of support from the family of climate change institutions, what official support climate migrants receive comes mostly from international humanitarian aid agencies and home country governments. International humanitarian aid, however, mostly supports IDP camps, refugee camps for people who cross national borders, and resettlement of persons accepted as refugees into advanced countries. IDP camps conventionally expect residents to return home when the weather disaster or conflict has abated, but many people displaced by climate change have no home to return to. Support for basic needs in IDP and refugee camps has fallen increasingly short as more people are displaced globally than at any time in history, as their numbers continue to rise, and as other crises such as COVID-19 and the war in Ukraine compete for resources [68]. Climate migrants in more vulnerable countries generally cannot rely on their home country governments to secure their basic needs. With no major international agency dedicated to supporting climate-displaced persons who move to find a new home, they are often left largely to their own devices.

Besides the human costs to migrants themselves, climate migration also imposes costs on host communities. We know that the overwhelming majority of climate migrants remain in their countries of origin [69], but we do not know how many cross borders, enter refugee camps, or seek employment in new home countries. Whether they relocate domestically or internationally, however, as their numbers increase, without adequate official support their presence is often politically destabilizing. Injustice due to neglect of polluters' responsibilities involves not only victimization of climate migrants themselves but also political conflict and other effects on receiving communities.

Climate migrants are not, of course, the only persons who experience great risk from climate change who are neglected by UNFCCC systems, but their neglect is particularly egregious. Although industrialized countries are largely responsible for the causes of climate migration, with most *de facto* institutional support for climate migrants coming from humanitarian aid agencies and from governments of more vulnerable countries, the public discourse surrounding migrants driven by climate change often fails to recognize industrialized countries' responsibility and the role of climate change as a driver.

Besides climate migrants being unable to press their governments to represent their interests in proposals for official adaptation support, their governments often have less influence over the substance of these proposals than the international development agencies involved in allocating funds and in implementing projects. Weak support for climate migrants' well-being is due not only to their own political weakness in their home countries but also to their home countries' weakness in the international arena. Resource competitions that begin with oil companies opposing developed countries' responsibility for GHG emissions and proceed through UNFCCC negotiations that reflect governments' relative geopolitical powers are finally worked out through bureaucratic processes that reflect the long history of competing interests in the international development community.

In the original position for climate justice, by contrast, the basic needs of climate migrants are considered side by side with other adaptation needs. Here it is apparent that the victimization climate migrants have already experienced is a great injustice and that the reduction of future risks requires major institutional reforms.

6. Conclusion: Stronger global governance is needed for climate justice

From the original position, we see that climate justice starts with dividing the 1003 Gt working carbon budget (at this writing) for limiting global warming to well below 2°C among the world's countries. Additional funds are needed from developed countries to help developing countries stay within their carbon budgets while in most cases continuing to industrialize and expand reliable access to electricity to their entire populations. Institutions for supporting developing countries' transitions to renewable energy need to be greatly strengthened. The UNFCCC needs to be empowered to tax developed countries for their fair share of costs and sanctions need to be devised to deter countries from exceeding their carbon budgets.

Just adaptation starts with assessing present and future risks to persons and to states from the effects of climate change and with strategies for reducing these risks. Although countries with high cumulative and current GHG emissions need to be held responsible for preventing significant deficits in basic needs, this does not excuse developing country governments and households from making their own prudent provisions. In principle, developed countries are to be held responsible for deficits that remain after reasonable adaptation efforts by developing country governments and their citizens or residents. If the burden of this support were to undermine the satisfaction of basic needs in advanced countries, levels of responsibility would be reduced. Since exceeding the global carbon budget would increase developed countries' burden of support for developing countries' adaptation, this gives developed countries an incentive to support the international carbon budget system.

Just adaptation clearly requires adaptation funding to be greatly increased and implementing institutions to be greatly strengthened. Assessing adaptation in terms of outcomes rather than levels of spending establishes incentives for enhancing the efficiency of adaptation spending and its associated institutions.

Implementing climate justice requires overcoming difficulties in defining terms, measuring key variables, building reasonably effective implementing institutions, assessing results, and applying stipulated sanctions (although hopefully, the threat of sanctions would be sufficient to secure compliance), not to mention securing developed countries' participation. The increase needed from the established $100 billion commitment could approach an order of magnitude. Considering that world military expenditures in 2023 were $2443 billion, about 2.3% of global gross domestic product (GDP) [70], the proposed reforms would probably require a reorientation in conceptions of national security as well as increased international cooperation.

If the UNFCCC were to secure powers to tax developed countries on the order of 0.5 to 1 percent of global GDP and to oversee fair adaptation, particularly in developing countries, it would constitute the strongest global government in history. While it would be in the service of climate justice, Rawls and other philosophers such as Immanuel Kant have been skeptical of global government, fearing that it would either become despotic or face constant rebellions on the periphery [7]. Although the powers envisaged fall far short of those held by national governments, such as holding a monopoly on legitimate violence within the national territory, constraints to its power and checks and balances among its constituent institutions and between it and national governments would need to be carefully devised and maintained. Losses entailed in national powers can only be justified by threats the climate crisis presents to people's liberty and autonomy, to international peace, and to the well-being of present and future generations.

The proposed reforms would also generate incentives for finding technical solutions for reducing atmospheric levels of GHGs more efficiently and for reducing harms from climate change, such as through geoengineering. Should relatively safe geoengineering "solutions" be found, a stronger UNFCCC would be better placed to secure the international agreement and global governance needed to legitimate their implementation.

While the IPCC calls for an "iterative risk management approach," we have seen that present institutions for climate governance cannot be expected to carry out such a process rationally or fairly. Effective risk management can only be achieved by significantly enhancing the powers of the UNFCCC and its associated institutions.

Author details

Paul Clements
Western Michigan University, Kalamazoo, USA

*Address all correspondence to: clements@wmich.edu

References

[1] Aramco. Aramco Announces Record Full-Year 2022 Results. 2023. Available from: https://www.aramco.com/en/news-media/news/2023/aramco-announces-full-year-2022-results [Accessed: June 13, 2024]

[2] Valle S. Exxon Smashes Western Oil Majors' Profits with $56 Billion in 2022. 2023. Available from: https://www.reuters.com/business/energy/exxon-smashes-western-oil-majors-earnings-record-with-59-billion-profit-2023-01-31/ [Accessed: June 13, 2024]

[3] World Bank Group. Pakistan: Flood Damages and Economic Losses Over USD 30 billion and Reconstruction Needs Over USD 16 billion – New Assessment. 2022. Available from: https://www.worldbank.org/en/news/press-release/2022/10/28/pakistan-flood-damages-and-economic-losses-over-usd-30-billion-and-reconstruction-needs-over-usd-16-billion-new-assessme [Accessed: June 13, 2024]

[4] International Organization for Migration. Displacement in Somalia Reaches Record High 3.8 Million: IOM Deputy Director Calls for Sustainable Solutions. 2023. Available from: https://www.iom.int/news/displacement-somalia-reaches-record-high-38-million-iom-deputy-director-general-calls-sustainable-solutions [Accessed: June 13, 2024]

[5] IPCC. Climate Change 2007: Synthesis Report Summary for Policymakers. 2007. Available from: https://www.ipcc.ch/report/ar4/syr/ [Accessed: July 20, 2024]

[6] Rawls J. A Theory of Justice. Revised ed. Cambridge: Harvard University Press; 1971, 1999

[7] Rawls J. The Law of Peoples. Cambridge: Harvard University Press; 1999

[8] Met Office. Reasons for Concern: Assessing Climate-Related Risk. Available from: https://www.metoffice.gov.uk/weather/climate-change/organisations-and-reports/reasons-for-concern-assessing-climate-related-risks [Accessed: July 21, 2024]

[9] Hansen J et al. Global warming in the pipeline. Oxford Open Climate Change. 2023;**3**(1):1-33. DOI: 10.1093/oxfclm/kgad008

[10] Bugden D. Denial and distrust: Explaining the partisan climate gap. Climatic Change. 2022;**170**(34):1-23. DOI: 0.1007/s10584-022-03321-2

[11] US Environmental Protection Agency. Climate Change Indicators: Atmospheric Concentrations of Greenhouse Gases. 2024. Available from: https://www.epa.gov/climate-indicators/climate-change-indicators-atmospheric-concentrations-greenhouse-gases#:~:text=Carbon%20dioxide%20concentrations%20have%20increased,is%20due%20to%-20human%20activities [Accessed: July 20, 2024]

[12] United Nations. Paris Agreement. 2015. Available from: https://unfccc.int/sites/default/files/english_paris_agreement.pdf

[13] IPCC. Synthesis Report of the IPCC Sixth Assessment Report (AR6). 2023. Available from: https://www.ipcc.ch/report/sixth-assessment-report-cycle/ [Accessed: July 24, 2024]

[14] Statista. Annual Carbon Dioxide (CO_2) Emissions Worldwide from 1940

to 2023. 2024. Available from: https://www.statista.com/statistics/276629/global-co2-emissions/ [Accessed: July 09, 2024]

[15] Toussaint P. Loss and damage, climate victims, and international climate law: Looking Back, looking forward. Transnational Environmental Law. 2024;**13**(1):134-159. DOI: 10.1017/S2047102523000237

[16] United Nations. United Nations Framework Convention on Climate Change. 1992. Available from: https://unfccc.int/files/essential_background/background_publications_htmlpdf/application/pdf/conveng.pdf [Accessed: July 24, 2024]

[17] United Nations. Addendum Part two: Actions Taken by the Conference of the Parties at its Twenty-First Session. FCCC/CP/2015/10/Add.1. Available from: https://www.un.org/en/development/desa/population/migration/generalassembly/docs/globalcompact/FCCC_CP_2015_10_Add.1.pdf [Accessed: July 24, 2024]

[18] Kramer RC. Carbon Criminals, Climate Crimes. New Brunswick: Rutgers University Press; 2020

[19] Congress.gov. S. Res. 98- A Resolution Expressing the Sense of the Senate Regarding Conditions for the United States Becoming a Signatory on Any International Agreement on Greenhouse Gas Emissions under the United Nations Framework Convention on Climate Change. 1997. Available from: https://www.congress.gov/bill/105th-congress/senate-resolution/98 [Accessed: July 24, 2024]

[20] Statista.com. Annual Carbon Dioxide (CO_2) Emissions Worldwide from 1940 to 2023. Available from: https://www.statista.com/statistics/276629/global-co2-emissions/ [Accessed: July 21, 2024]

[21] Faulkner R. The Paris agreement and the new logic of international climate politics. International Affairs. 2016;**92**(5):1107-1125. DOI: 10.1111/1468-2346.12708

[22] Hope B. Top 10: Nations That Are Leading the Renewable Energy Charge. 2022. Available from: https://sustainabilitymag.com/net-zero/top-10-nations-that-are-leading-the-renewable-energy-charge-countries-emissions [Accessed: July 22, 2024]

[23] International Energy Agency. Net Zero by 2050: A Roadmap for the Global Energy Sector. 2021. Available from: https://www.iea.org/reports/net-zero-by-2050 [Accessed: July 24, 2024]

[24] Ballester J et al. Heat-related mortality in Europe during the summer of 2022. Nature Medicine. 2023;**29**:1857-1866. DOI: 10.1038/s41591-023-02419-z

[25] de Bont J et al. Impact of heatwaves on all-cause mortality in India: A comprehensive multi-city study. Environment International. 2024;**184**(108461):1-10. DOI: 10.1016/j.envint.2024.108461

[26] Arusu S. India is Likely Undercounting Heat Deaths, Affecting its Response to Increasingly Harsh Heat Waves. 2024. AP Climate. Available from: https://apnews.com/article/india-heat-wave-death-toll-undercounting-climate-change-f54464851e45fbc4019caededa90ce12 [Accessed: July 20, 2024]

[27] Ahmed B. Climate migrants in Bangladesh: A journey towards uncertainty! In: Global Shifts Colloquium on Seeking Refuge in the Climate Emergency. Philadelphia, PA, USA:

UCL Discovery; 2020. pp. 1-7. Available from: https://discovery.ucl.ac.uk/id/eprint/10137803/

[28] Khan MA. Livelihood WASH related hardships and needs assessment of climate migrants: Evidence from urban slums in Bangladesh. Heliyon. 2022;**8**(5) e09355:1-9. DOI: 10.1016/j.heliyon.2022.e09355

[29] Adri N, Simon D. A tale of two groups: Focusing on the differential vulnerability of 'climate-induced' and 'non-climate-induced' migrants in Dhaka City. Climate and Development. 2018;**10**(4):321-336. DOI: 10.1080/17565529.2017.1291402

[30] Li D, Yuan J, Copp RE. Escalating global exposure to compound heat-humidity extremes with warming. Environmental Research Letters. 2020;**15**(6):1-11. DOI: 10.1088/1748-9326/ab7d04

[31] Statista. Number of Refugees and Internally Displaced Persons (IDPs) Worldwide from 2000 to 2022. 2024. Available from: https://www.statista.com/statistics/268702/number-of-refugees-and-internally-displaced-persons-worldwide-since-2000/ [Accessed: July 07, 2024]

[32] Bilal A, Känzig D R. The Macroeconomic Impact of Climate Change: Global vs. Local Temperature. 2024. National Bureau of Economic Research Working Paper 32450. Available from: http://www.nber.org/papers/w32450 [Accessed: July 24, 2024]

[33] Popovich N, Plumer B. Who has the Most historical responsibility for climate change? The New York Times. 12 Nov 2021. Available from: https://www.nytimes.com/interactive/2021/11/12/climate/cop26-emissions-compensation.html [Accessed: July 08, 2024]

[34] Mellin C et al. Safeguarding nutrients from coral reefs under climate change. Nature Ecology and Evolution. 2022;**6**:1808-1817. DOI: 10.1038/s41559-022-01878-w

[35] Wunderling N et al. Recurrent droughts increase risk of cascading tipping events by outpacing adaptive capacities in the Amazon rainforest. PNAS Environmental Sciences. 2022;**119**(32):1-11. DOI: 10.1073/pnas.2120777119

[36] IPCC. Climate Change 2021: The Physical Science Basis. 2021. Available from: https://www.ipcc.ch/report/ar6/wg1/ [Accessed: July 23, 2024]

[37] Carter PD. Implications for Biodiversity of Potentially Committed Global Climate Change (from Science and Policy). In: Filho WL, Barbir J, Preziosi R, editors. Handbook of Climate Change and Biodiversity. Cham: Springer; 2019. pp. 135-149

[38] Kulp SA, Strauss BH. New elevation data triple estimates of global vulnerability to sea-level rise and coastal flooding. Nature Communications. 2019;**10**(4844):1-12. DOI: 10.1038/s41467-019-12808-z

[39] Minderhoud PSJ et al. Mekong delta much lower than previously assumed in sea-level rise impact assessments. Nature Communications. 2019;**10**(3847):1-13. DOI: 10.1038/s41467-019-11602-1

[40] Hansen J et al. Ice melt, sea level rise, and superstorms: Evidence from paleoclimate data, climate modeling, and modern observations that 2°C global warming could be dangerous. Atmospheric Chemistry and Physics. 2016;**16**:3761-3812. DOI: 10.5194/acp-16-3761-2016

[41] UN Environment Programme. Emissions Gap Report 2023. Broken

Record: Temperatures Hit New Highs, Yet World Fails to Cut Emissions (again). 2023. Available from: https://www.unep.org/resources/emissions-gap-report-2023 [Accessed: July 23, 2024]

[42] Pauw WP et al. Conditional nationally determined contributions in the Paris agreement: Foothold for equity or Achilles heel? Climate Policy. 2020;**20**(4):468-484. DOI: 10.1080/14693062.2019.1635874

[43] US Department of State. On the U.S. Withdrawal from the Paris Agreement. 2019. Available from:https://2017-2021.state.gov/on-the-u-s-withdrawal-from-the-paris-agreement/ [Accessed: July 09, 2024]

[44] Popovich N, Albeck-Ripka L, Pierre-Louis, K. The Trump Administration Rolled Back More than 100 Environmental Rules. Here's the Full List. The New York Times. 2020. Available from: https://www.nytimes.com/interactive/2020/climate/trump-environment-rollbacks-list.html [Accessed: July 09, 2024]

[45] Tirumurti TS. Backtracking on climate action by the developed countries: Some reflections of a negotiator. Environmental Policy and Law. 2022;**52**:463-471. DOI: 10.3233/EPL-219046

[46] Oxfam International. The True Value of Climate Finance Is a Third of What Developed Countries Report. 2022. Available from: https://www.oxfam.org/en/press-releases/true-value-climate-finance-third-what-developed-countries-report-oxfam [Accessed: July 10, 2024]

[47] Global Environment Facility. Promoting Low-carbon Electric Bus Transport in Mauritius. UNDP Project Proposal submitted 10/8/2019. 2022. Available from:https://www.thegef.org/sites/default/files/web-documents/10372_CC_PIF.pdf [Accessed: July 23, 2024]

[48] African Development Fund. Ghana Mini Grid and Solar PV Net Metering Project Appraisal Report. 2021. Available from: https://www.afdb.org/en/documents/ghana-ghana-mini-grid-and-solar-pv-net-metering-project-appraisal-report [Accessed: July 20, 2024]

[49] Berwyn B. Average Global Temperature Has Warmed 1.5 Degrees Celsius Above Pre-industrial levels for 12 Months in a Row. Inside Climate News. 2024. Available from: https://insideclimatenews.org/news/09072024/average-global-temperatures-above-pre-industrial-levels-for-12-months/ [Accessed: July 11, 2024]

[50] Kenehan S. In defense of the duty to assist: A response to critics on the viability of a Rawlsian approach to climate change. Critical Review of International Social and Political Philosophy. 2015;**18**(3):308-327. DOI: 10.1080/13698230.2013.839428

[51] UN Environment Programme. Underfinanced. Underprepared. Inadequate Investment and Planning on Climate Adaptation Leaves World Exposed. Adaptation Gap Report 2023. Available from: https://www.unep.org/resources/adaptation-gap-report-2023#:~:text=UNEP's%20Adaptation%20Gap%20Report%202023,these%20rising%20climate%20change%20impacts [Accessed: July 23, 2024]

[52] Global Commission on Adaptation. Adapt Now: A Global Call for Leadership on Climate Resilience. Rotterdam and Washington, DC: Global Center for Adaptation and World Resources Institute; 2019. Available from: https://gca.org/reports/

adapt-now-a-global-call-for-leadership-on-climate-resilience/ [Accessed: July 23, 2024]

[53] Buchner B et al. Global Landscape of Climate Finance. 2023. Available from: https://www.climatepolicyinitiative.org/publication/global-landscape-of-climate-finance-2023/ [Accessed: July 23, 2024]

[54] Pörtner H et al. Climate Change 2022: Impacts, Adaptation and Vulnerability—Summary for Policymakers, Working Group II contribution to the Sixth Assessment Report of the Intergovernmental Panel on Climate Change. 2022. Available from: https://www.ipcc.ch/report/ar6/wg2/ [Accessed: July 23, 2024]

[55] Woodruff SC, Regan P. Quality of national adaptation plans and opportunities for improvements. Mitigation and Adaptation Strategies for Global Change. 2019;**24**:53-71. DOI: 10.1007/s11027-018-9794-z

[56] Garshagen M, Doshi D. Does funds-based adaptation finance reach the most vulnerable countries? Global Environmental Change. 2022;**73**(102450):1-10. DOI: 10.1016/j.gloenvcha.2021.102450

[57] Kabir A et al. Determinants of climate change adaptation strategies in the coastal zone of Bangladesh: Implications for adaptation to climate change in developing countries. Mitigation and Adaptation Strategies for Global Change. 2021;**26**(30):1-25. DOI: 10.1007/s11027-021-09968-z

[58] Tendler J. Inside Foreign Aid. Baltimore: Johns Hopkins University Press; 1975

[59] Islam MM. Distributive justice and global climate finance – Recipients climate vulnerability and the allocation of climate funds. Global Environmental Change. 2022;**73**:102475. DOI: 10.1016/j.gloenvcha.2022.102475

[60] Eriksen S et al. Adaptation interventions and their effect on vulnerability in developing countries: Help, hindrance or irrelevance? World Development. 2021;**141**:105383. DOI: 10.1016/j.worlddev.2020.105383

[61] Pisor AC et al. Effective climate change adaptation means supporting community autonomy. Nature Climate Change. 2022;**12**:213-215. DOI: 10.1038/s41558-022-01303-x

[62] United Nations Climate Change. Task Force on Displacement. 2024. Available from: https://unfccc.int/process/bodies/constituted-bodies/WIMExCom/TFD [Accessed: July 16, 2024]

[63] United Nations Climate Change. NAPA Priorities Database. 2024. Available from: https://unfccc.int/topics/resilience/workstreams/national-adaptation-programmes-of-action/napa-background [Accessed: July 16, 2024]

[64] Adaptation Fund. Projects Table View. 2024. Available from: https://www.adaptation-fund.org/projects-programmes/project-information/projects-table-view/ [Accessed: July 16, 2024]

[65] GEF. Project Database. 2024. Available from: https://www.thegef.org/projects-operations/database?f%5B0%5D=focal_areas%3A2207 [Accessed: July 16, 2024]

[66] UNFCCC. Operationalization of the New Funding Arrangements for Responding to Loss and Damage and the Fund Established in Paragraph 3 of Decisions 2/CP.27 and 2/CMA.4. FCCC/CP/2023/9–FCCC/PA/CMA/2023/9. 2023. Available from: https://unfccc.int/sites/default/files/resource/

cp2023_09_cma2023_09.pdf [Accessed: July 20, 2024]

[67] Schalatek L, Richards J. The Loss and Damage Fund Board: Getting it Right from the Start. 2024. Available from: https://us.boell.org/en/2024/03/18/loss-and-damage-fund-board-getting-it-right-start#:~:text=As%20of%20March%202024%2C%20USD,States%20of%20America%20whose%20pledge [Accessed: January 22, 2024]

[68] UNHCR. UNHCR Warns Rising Tide of Hunger, Insecurity and Underfunding Worsening Gender-Based Violence Risks. 2022. Available from: https://www.unhcr.org/us/news/press-releases/unhcr-warns-rising-tide-hunger-insecurity-and-underfunding-worsening-gender [Accessed: July 22, 2024]

[69] Gemenne F et al. Forced Displacement Related to the Impacts of Climate Change and Disasters. 2021. Available from: https://www.unhcr.org/people-forced-to-flee-book/wp-content/uploads/sites/137/2021/10/Franc%CC%A7ois-Gemenne-et-al_Forced-displacement-related-to-the-impacts-of-climate-change-and-disasters.pdf [Accessed: July 22, 2024]

[70] Tian N et al. Trends in World Military Expenditure, 2023. 2024. Available from: https://www.sipri.org/sites/default/files/2024-04/2404_fs_milex_2023.pdf [Accessed: July 19, 2024]

Chapter 6

Enhancing Climate Resilience in Cabo Verde: Strategies, Challenges, and Opportunities

Pedro Andrade Matos

Abstract

Climate change presents significant challenges for Cabo Verde, especially impacting its agriculture sector, which is vital for the country's economy. With an arid climate and limited land, Cabo Verde faces issues, such as soil degradation, drought, and a heavy reliance on food imports. These challenges are compounded by socio-economic factors, including widespread poverty and inequality, which exacerbate vulnerability and limit effective disaster risk management. Efforts to adapt to climate change include implementing soil and water conservation measures, such as terraces and drip irrigation, and diversifying income sources through remittances. A critical aspect of building resilience is integrating climate considerations into all public policy agendas. This involves leveraging traditional knowledge, promoting climate-smart technologies, and emphasizing gender equality to ensure that women, who play a crucial role in agriculture, are recognized as agents of transformation rather than just victims of climate change. Strengthening internal capacities and fostering international cooperation are essential for Cabo Verde to effectively address climate challenges. By adopting an integrated approach that combines technological innovation, community engagement, and strategic policy development, the country can turn climate challenges into opportunities and prosperity for sustainable growth and development.

Keywords: agriculture, Cabo Verde, climate change, resilience, opportunities

1. Introduction

Ensuring the sustainability of island states amidst high vulnerability and multiple crises presents a significant challenge. Factors, such as high food prices, conflicts, and market distortions, create a cycle of vulnerability and instability, impeding efforts to establish sustainable development pathways worldwide [1].

Climate change exacerbates these challenges for Small Island Developing States (SIDS), such as Cabo Verde, affecting various aspects of life and development. With its arid climate, limited land resources, and heavy reliance on food imports, Cabo Verde faces unique vulnerabilities intensified by climate variability and extreme weather events.

Agriculture, a vital sector for Cabo Verde's economy, is undermined by low technology adoption, land degradation, and climate change impacts. Despite efforts in soil and water conservation, issues such as drought, erosion, and soil fertility loss persist. The country's reliance on food imports and the economic repercussions of global crises highlight the need for a robust agricultural strategy.

Socio-economic factors further complicate the situation as elevated levels of poverty and inequality restrict the ability of vulnerable populations to manage disaster risks effectively. Rapid urbanization and the lack of coordinated disaster risk management strategies exacerbate these challenges.

This article aims to explore how Cabo Verde can enhance its climate change adaptation strategies by integrating local policies and practical approaches to strengthen agricultural and institutional resilience.

By leveraging traditional knowledge, advancing technological innovations, and emphasizing gender equality, Cabo Verde can turn its climate challenges into opportunities for sustainable growth. Building internal capacities and fostering international cooperation are crucial for long-term resilience and effectively navigating climate change complexities.

The research employs a multifaceted approach, including a comprehensive review of existing studies, evaluations of national and local policies, and assessments of community involvement and traditional practices. The climate projection for Cape Verde is based on data compiled from global climate models from the Coupled Model Intercomparison Project, supervised by the World Climate Research Program (Coupled Model Intercomparison Project Phase 6), considering different emission scenarios or shared socio-economic pathways (SSPs). The data are generated from the Climate Change Knowledge Portal.

The article is organized into six sections. The first section addresses the key concepts related to climate change and discusses the challenges faced by Small Island Developing States, focusing on Cabo Verde, in terms of mitigation and adaptation. The second section provides an overview of climate geography, situating Cabo Verde in a region highly sensitive to climate change. The third section analyzes the country's resilience from a multilevel perspective, examining how Cabo Verde is preparing, with various capacities, to face climate risks. The fourth section focuses on the resilience of the agricultural sector as it is the most affected by climate change. The fifth section discusses the challenges and opportunities that climate change presents for the country. Finally, the sixth section interprets the risks and uncertainties related to climate, exploring different future scenarios.

2. Concepts and literature review

Climate change is related to extreme weather events, such as rising global temperatures, storms, tropical cyclones, sea-level rise, and hurricanes. The term "refers to any change in climate over time due to natural variability or resulting from human activity" ([2], p. 6). Human activities in conflict with the environment have caused, directly or indirectly, extreme and intense weather events, increasing the risks of environmental disasters.

Disasters involve human presence in their course, that is, disasters are not natural phenomena, but rather social constructions resulting from the conflicting interaction between society and the environment [3]. A disaster corresponds to a serious disruption in the functioning of a community or society, involving widespread human, material, economic, or environmental losses and impacts, which exceed the ability of the affected community or society to cope using its own resources [4, 5].

Disaster impacts may include loss of life, injury, disease, and other negative effects on human physical, mental, and social well-being, along with damage to property, destruction of assets, loss of services, social and economic disruption, and environmental degradation [5]. However, it is necessary to differentiate between disaster risk reduction and climate change adaptation (CCA). While DRR is relevant for all types of risks, CCA is more specifically oriented toward climate-related risks. DRR is more concerned with the present, assessing and addressing existing risks, while CCA focuses on the future, dealing with uncertainties and new risks ([6], p. 251).

Climate and disaster resilient development (or Climate and Disaster Resilience) comprise a set of institutional arrangements, processes, and instruments that help identify the risks from disasters, climate extremes, gradual and long-term climatic changes, and their associated impacts and the design of measures to reduce, transfer, and prepare for such risks. Climate and disaster resilient development combine development benefits with reductions in vulnerability over the short and long term using a development planning, multi-sectoral, and multi-stakeholder approach [7].

The impact of a natural disaster varies, depending on the magnitude of the adverse event (threat or danger) and the degree of vulnerability of the affected system. Vulnerability refers to the degree of loss expected for an element or set of elements in a situation of danger, that is, exposed to a specific threat, as a result of physical, social, economic, and environmental factors ([2], p. 6).

Vulnerability encompasses a variety of concepts and elements, including sensitivity or susceptibility to harm and a lack of capacity to cope and adapt [8]. The assessment of this capacity can be framed by resilience, which is decisive during the manifestation of a natural or socio-environmental event and post-event reconstruction. The term refers to the capacity of social, economic, and environmental systems to cope with a hazardous event or trend or disturbance, responding or reorganizing in ways that maintain their essential function, identity, and structure while also maintaining the capacity for adaptation, learning, and transformation [8]. Douglas Sono et al. [9] define climate resilience as the ability of a country or region to resist, absorb, recover, adapt, and develop through continuous learning and adjusting internal economic, social, environmental, and infrastructure systems in the face of climate change-related risks caused by acute shocks or chronic stresses. Resilience encompasses systems, such as food security, infrastructure, and governance [9]. Resilience is not given; it is created and manifests itself in adverse contexts.

This matters as it highlights that capacities must be built for the country to withstand the effects of climate change. Cabo Verde follows the IPCC definition, describing resilience as "the ability of a social or ecological system to absorb disturbances while retaining the same basic structure and ways of functioning, the capacity for self-organization, and the capacity to adapt to stress and change" [10].

State capacities are essential for the provision of public goods, including environmental protection [11, 12]. In fact, sensitive issues such as climate change test the very capacity of states, especially in rich countries, to develop policies against the opposition of powerful interest groups, in view of the multiple and multilevel administrative effects of the issue, which must be considered by institutions for the implementation of policies and management of incompatible interests.

The research produced by Sono et al. [9] indicates that countries with effective governance, such as Botswana, Mauritius, and Cabo Verde, demonstrate higher resilience to climate change impacts due to better control over corruption and accountability in implementing climate programs. In contrast, countries, such as Somalia and South Sudan, show lower resilience due to weak governance structures.

When considering these elements, the category of countries that stands out is SIDS. Vulnerability is related to the "ability of SIDS to adjust to the risks arising from climate change" [13]. SIDS face several challenges, including sea-level rise, changing precipitation patterns, and increased intensity of tropical storms and cyclones [14].

The acronym SIDS designates a group of island countries that face similar challenges and objectives regarding sustainable development and environmental issues, especially climate change. In general, these states have a small territorial area, and their special economic zones are larger than their land area.

They are vulnerable to natural weather events, such as storms, cyclones, and hurricanes, and face climate variations, which result in droughts and floods due to changes in the ocean circulation system that alter precipitation patterns. They are dependent on maritime resources and are far from the main markets, which affects their trade competitiveness. They are also subject to external shocks. In short, they need international cooperation to build and strengthen national capacities in relation to climate change [5, 13–15].

As small island developing states (SIDS) navigate shifts in hurricane frequency, rainfall patterns, and drought conditions—particularly in coastal zones and within the water and agriculture sectors—they encounter significant challenges, with financing emerging as a major constraint [16]. Adaptation strategies in SIDS often encompass structural, physical, and behavioral adjustments. However, there is a notable lack of comprehensive evaluation regarding the effectiveness of these measures [14]. The vulnerabilities of these countries manifest in resource insecurity, significant imports, poor waste management, and an inability to develop economies of scale [17].

Bahers' [17] proposal is to introduce the concept of "sociometabolic vulnerability," which provides links between patterns of resource use, systemic risks, and vulnerability. Studying the island of Comoros, they conclude that there is a very low level of resource use and heavy dependence on critical imports that cover large distances, which are vulnerable to price and climate shocks. In this sense, they suggest that informal activities in resource extraction play a significant role in building resilience and the development of communities. Despite these obstacles, many SIDS have strengthened their resilience by leveraging traditional knowledge and pioneering innovative mitigation and adaptation strategies [18]. Unfortunately, research on adaptation to SIDS remains fragmented, with a predominant focus on the Pacific region and individual island nations [14].

Studying Cabo Verde's climate change allows for understanding its impacts on biodiversity, energy systems, and social structures, as well as developing effective conservation and adaptation strategies for this vulnerable small country. Indeed, climate change poses significant challenges for Cabo Verde, affecting biodiversity, energy systems, and socio-economic factors [19]. Long-term climate trends indicate increasing temperatures and decreasing precipitation, signaling ongoing aridification processes [20]. Despite renewable energy potential, Cabo Verde heavily relies on fossil fuels for electricity generation [21]. Addressing climate change requires consideration of gender dynamics as women and men have distinct roles and responsibilities concerning the environment. Promoting gender equality may enhance the effectiveness of mitigation and adaptation strategies [22].

3. Cape Verde: Geography of climate change

Cabo Verde is an archipelago located in the Atlantic Ocean, off the coasts of Senegal and Mauritania, about 455 km from the west coast of Africa (**Figure 1**). The archipelago encompasses ten islands (nine inhabited: Santo Antão, São Vicente, São Nicolau, Sal, Boa Vista, Maio, Santiago, Fogo, and Brava; and one uninhabited: Santa Luzia) and eight islets (Branco, Raso, Grande, Luís Carneiro, Cima, and others). Since 1990, the annual temperature in Cabo Verde has increased by around 0.04%, and projections indicate an increase of about 1% for the period 2011–2040 and 3% by the end of the century [10].

The country's climate is considered mild and tropical, with a strong influence from the cold Canary Current, which affects temperature. Throughout the year, the Azores anticyclones also influence this climate. The islands with an arid climate are Maio, São Vicente, and Brava, while Santo Antão and Santiago have semiarid and arid climates. On the other hand, only Fogo Island has a subhumid climate, as well as arid and semiarid regions [24].

Due to its geographical location and climate system, Cabo Verde is vulnerable to climate change, with consequences seen in terms of climate aridity, frequent droughts, soil degradation, and loss of biodiversity. In this context, Cabo Verde is different from other SIDS. In addition to being a highly vulnerable country, it is located in the Sahel, a region that faces enormous climatic challenges, including conflicts, environmental degradation, food insecurity, water scarcity, drought, desertification, and socio-economic disturbances, such as uncontrolled urbanization, power fluctuations, a weak financial system, unemployment, weak institutions, and mismanagement of public funds [1, 9].

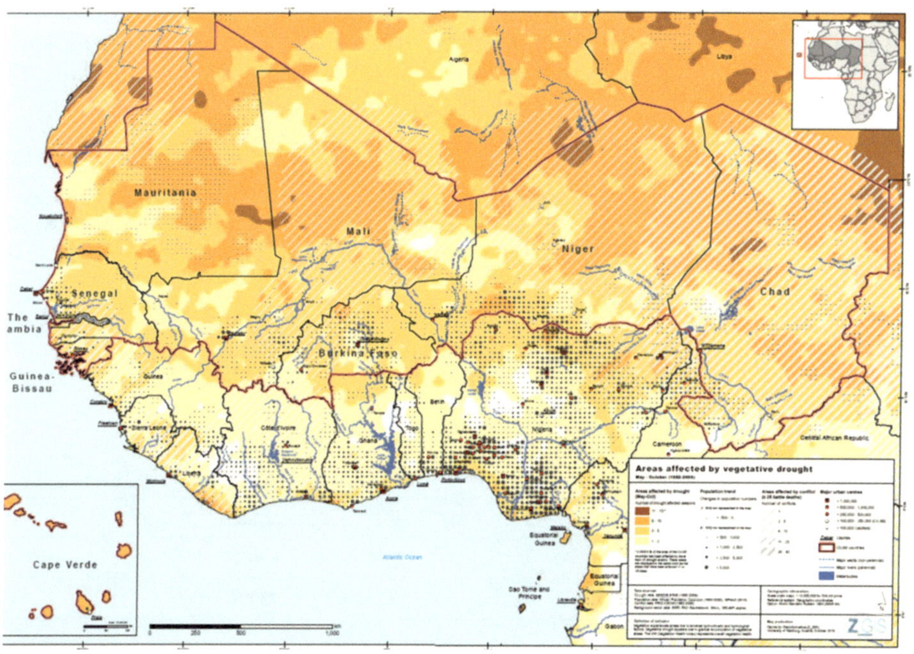

Figure 1.
Sahel drought map – Cabo Verde. Source: MrGeogWagg [23].

Cabo Verde has a history of severe droughts and famines that devastated the population. However, after gaining political independence in 1975, the country addressed poverty. Independence also provided an opportunity to combat famine and drought on the islands (**Table 1**).

As a result of successive government policies, Cabo Verde is now recognized as a success story in transforming a harsh environment into a more resilient one that can better withstand extreme drought periods without resulting in widespread famine and starvation. Cabo Verde is a prominent case in the sub-Saharan African region: "The vulnerability and readiness metric suggested that Cabo Verde is the only country in SSA to have low vulnerability and high readiness, while most countries have high vulnerability and low readiness, making them the least resilient countries needing urgent mitigation and adaptation actions" [9].

This does not mean that hunger has been eliminated. Currently, the technical term "food security" is often used to conceal a reality in which the echoes of the poorest families are confirmed by the lack of food and the quality of food necessary to have a decent meal. Drought, as a structuring and shaping factor of Cabo Verdean society, has not yet been addressed efficiently and produces several negative effects on society, especially the departure of young people from their communities. The situation has worsened with climate change, making agricultural production deficits and crop uncertainties more frequent each year (**Figure 2**).

Drought is, in fact, the type of disaster that has resulted in the most deaths and affected the most people. In this sense, the country's climate resilience must consider drought as a structuring element. However, the persistent drought highlights the failure to eliminate a set of barriers to socio-economic development.

Climate change is exacerbating drought conditions in Cabo Verde, posing significant challenges to agriculture and food security [1, 26]. The archipelago, characterized by limited resources and vulnerability to natural disasters, experiences regular extreme drought conditions that particularly impact the agricultural sector [1].

To mitigate water scarcity, the government has implemented strategies, such as small-scale irrigation and drip irrigation technology [26]. A pilot project demonstrated that the safe and profitable reuse of treated water through subsurface drip

Disaster type	Disaster subtype	Events count	Total deaths	Total affected	Total damage ('000 US$)
Drought	Drought	10	85,000	40,000	—
Flood	Riverine flood	1	3	150	—
Storm	Tropical cyclone	3	41	7722	4100
Insect infestation	Locust	2	0	—	—
Epidemic	Bacterial disease	1	245	12,344	—
Epidemic	Viral disease	1	6	20,147	—
Volcanic activity	Ashfall	1	0	6306	—
Volcanic activity	Lava flow	1	0	2500	—

Source: Costa [1], p. 256.

Table 1.
Types of natural disasters and their impacts, Cabo Verde.

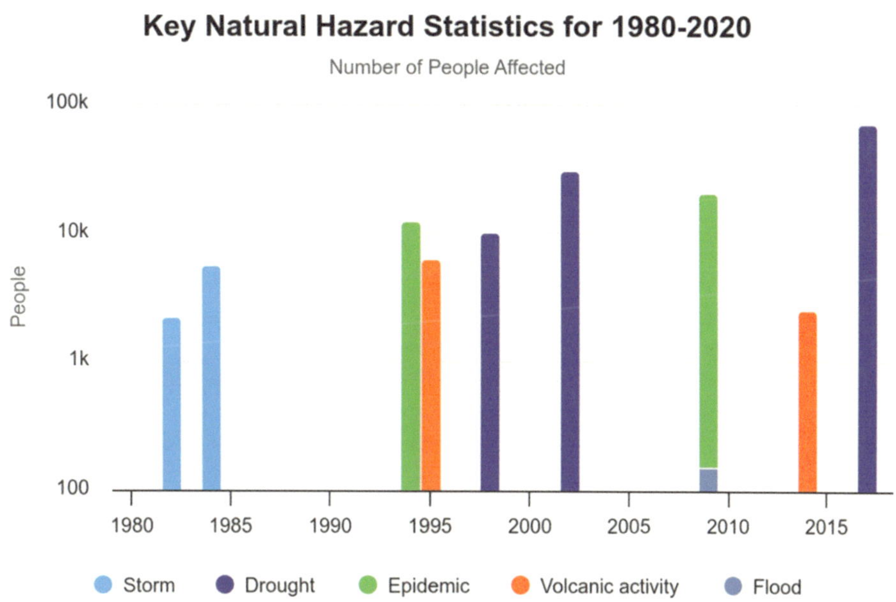

Figure 2.
Key natural hazard statistics for 1980–2020. Source: World Bank Climate Change Knowledge Portal [25].

irrigation (SDI) could improve water use efficiency and increase food production [27]. However, challenges remain, including the need for farmer training and addressing installation failures [27].

Studying resilience in the agricultural sector reveals two important characteristics: first, it is the sector most impacted by climate change; second, it corresponds to the sector where the most vulnerable social groups are working, especially women. Therefore, the analysis of this sector allows for the construction of robust strategies to guide decision-makers in more efficient policy formulation and public resource management.

4. Multilevel resilience: Cabo Verde's analysis

Resilience operates at multiple levels (individual, community, and national). The resilience at the individual level is influenced by community resilience, which, in turn, affects the national level. However, each level retains different characteristics of resilience [9]. At the local level, resilience is related to interventions designed to inform programming. To assess this level of resilience, participatory tools are used to gather information about living conditions.

The Pan-African research network Afrobarometer conducted an interview in 2021 with 1200 adult Cabo Verdeans to capture their perspectives, especially those of young people, on the performance of democracy and governance in the country. The results revealed that 58% of the interviewed population considered Cabo Verde to be a complete democracy or a democracy with minor problems, while 41% thought it was a democracy with major problems. Notably, 46% of the youth segment felt that "we are not a democracy, or we are a democracy with major problems," with unemployment being one of the main issues identified (71%). Crime and security were also highlighted as problems needing resolution [28].

Regarding climate change, Afrobarometer research conducted by Margaret Eduonoo [29] indicates that Cabo Verdean society is aware of "climate change," particularly due to drought as a shaping element. However, eight out of ten interviewees believe that the drought has become more severe in the past decade, worsening their quality of life [29].

At the community level, resilience includes factors, such as "the quality of environmental and natural resources management institutions, access to communal resources, quality of protective infrastructure, levels of peace and security, availability of contingency resources or social safety nets, and social participation" ([9], p. 4). Indicators at this level include income level, access to food, and access to basic services, such as health, assets, and social safety nets.

According to the "Leaving No One Behind" assessment conducted by the Cabo Verdean government in 2021 with support from the United Nations, "in 2020, 31.6% of the population lived below the national poverty line (less than US$ 2.77 per day), totaling 175,844 poor individuals. Poverty incidence varies between municipalities, with São Filipe at 65.8%, Santa Cruz at 61.7%, Tarrafal de São Nicolau at 54.5%, and São Lourenço dos Órgãos at 51.6%." The FAO [30] reports that the prevalence of undernourishment was 12.6% of the population between 2020 and 2022.

Statistics on household living conditions show that 73.8% of households had access to piped water connected to the public distribution network [31]. Most households use piped water (71.3%), while 11.8% rely on autotanks, 7.0% on neighbors' houses, 6.1% on fountains, and 3.7% use other protected or non-protected sources (boreholes, wells, levadas, springs, etc.). The majority of the rural population (58.1%) uses other sources of water [31].

Women are responsible for transporting water, averaging 21.2 minutes to collect and return it [31]. This division of labor impacts individual and community-level resilience. Climate change exacerbates global water insecurity, disproportionately affecting women [32]. Women face greater constraints in adapting to climate change due to limited access to resources and information [33]. In Nepal, decreasing water availability has increased women's workload for water collection, leading to longer distances and more time spent [34, 35]. This time poverty affects the well-being of women and their families [34].

The impacts of climate change on water resources are often perceived as less urgent compared to immediate problems, highlighting the need for sensitization and awareness at individual, household, and community levels [35]. Addressing these challenges requires intersectional analyses and the promotion of gender-responsive local adaptation planning in water management [32, 34].

Resilience at the national level builds upon the first two levels and focuses on systems, institutions, and policies that contribute to national resilience. A key variable in this context is the national government's capacity to respond to climate-related disasters ([9], p. 4).

Resilience is essential for achieving sustainable development in small island developing states (SIDS). However, available financing for this purpose is limited and difficult to access [36]. The degree of vulnerability to climate change is directly linked to state capacities, revealing challenges in accessing basic human rights, education, health services, security, and governance [37, 38]. Therefore, it is crucial to develop capacities to mobilize international resources and partnerships essential for climate finance related to adaptation policies and mitigation measures [13].

To enhance institutional capacity in response to climate change challenges, including the growing need for domestic responses and new climate financing modalities, the new climate governance framework was approved through Resolution No. 38/2024 of 10 May. This framework aims to facilitate the implementation of the nationally determined contributions (NDCs) and the National Adaptation Plan, as well as to optimize the use of available climate finance opportunities. It contributes to the strategic plan for sustainable development by focusing on inclusive processes, institutional coherence, and scientific excellence, enabling the operationalization of the enhanced transparency framework (Resolution No. 38/2024 of 10 May).

The climate governance framework encompasses eight main functions: (a) Decision-making; (b) Coordination; (c) Climate negotiation; (d) Knowledge management and early warning system; (e) Mobilization and management of climate support; (f) Communication and awareness; (g) Implementation; and (h) Transparency, monitoring, and evaluation.

Decision-making and coordination functions are carried out by the following bodies: (a) Interministerial Council for Climate Action (CIAC); (b) National Council for the Environment and Climate Action (CNAAC); (c) National Secretariat for Climate Action (SNAC).

The Interministerial Council for Climate Action is an interministerial decision-making and coordination body on climate policy and sectoral policies impacting national climate objectives, chaired by the Prime Minister. Members include government officials responsible for finance, social inclusion and development, territorial cohesion, foreign affairs and cooperation, education and science, health, tourism and transport, sea, agriculture, climate action and environment, water and energy, and spatial planning.

The National Council for the Environment and Climate Action is an advisory body that collaborates with the government department responsible for environment and climate action. Its fundamental tasks are to issue opinions and ensure the alignment of political and social positions regarding the integrated and sustainable management of the Environment in Cabo Verde, in relation to the national poverty reduction policy and economic growth.

The CNAAC includes representatives from government departments responsible for finance, economy and innovation, territorial cohesion, foreign affairs and cooperation, civil protection, education and science, health, tourism and transport, sea, agriculture and environment, energy, and spatial planning, as well as representatives from municipalities, civil society organizations, the private sector, and academia.

The National Secretariat for Climate Action (SNAC) is an executive body under the direct responsibility of the government member in charge of the climate change sector. It coordinates transparency and implementation, knowledge management and alert systems, climate negotiation, and communication and education.

The National Plan for Adaptation to Climate Change (PANA) has three objectives: (1) To create an enabling environment for integrating adaptation into planning and budgeting; (2) To improve the capacity to manage and share data and access technology and finance for adaptation; (3) To implement adaptation actions for greater resilience of the most vulnerable.

Cabo Verde recognizes its low institutional capacity and fragile mechanisms for coordination, partnership, and dialog on climate change issues [10]. Despite this, the country has relatively high institutional resilience [9] compared to other sub-Saharan African countries.

Among the Cabo Verdean population interviewed, the responsibility for combating climate change is perceived as falling primarily on developed countries (32%), followed by the government (29%) and civil society (23%). Only 34% of respondents believe the government is effectively addressing climate change. Additionally, 63% of respondents feel the government must act immediately to limit climate change, even if it incurs significant costs [29].

Climate change poses significant challenges to agriculture, necessitating effective institutional responses. Public institutions play a crucial role in building agricultural resilience by providing information and technology support to farmers [39]. However, these institutions often face capacity constraints, particularly in terms of financial, physical, and human resources [39]. While formal institutions are vital for developing place-based adaptation strategies, they must also acknowledge cultural factors and collaborate with informal institutions to be more effective [40].

In the context of digital technologies, climate-smart agriculture requires inclusive institutions that provide information, enable innovation, encourage investment, and offer insurance [41]. Although institutional support for adaptive capacity in some regions may appear adequate, assessing the success of these efforts remains challenging due to ongoing declines in agricultural system viability [42]. Improving institutional capacities and fostering collaboration between formal and informal institutions are critical for enhancing agricultural resilience to climate change.

5. Resilience strategies in agriculture

The literature analyzed by Klöck & Nunn [14] identifies a variety of adaptation strategies for small island developing states (SIDS), which can be categorized into key areas:

1. *Structural or physical changes*: This includes modifications to infrastructure to withstand climate impacts, such as building sea walls, improving drainage systems, and enhancing coastal defenses.

2. *Behavioral changes*: Communities often adapt by altering their daily practices in response to climate variability. This includes preparing for extreme weather events by preserving food, securing homes, and modifying agricultural practices (e.g., switching to different crops or fish species) to cope with changing environmental conditions [43, 44].

3. *Social adaptation*: This involves community-level initiatives, such as capacity building, awareness raising, and educational measures, aimed at enhancing resilience to climate impacts. Social networks play a crucial role in sharing knowledge and resources [45].

4. *Institutional adaptation:* This encompasses the development of laws, regulations, and policies that address a broader range of climate change impacts and promote sustainable practices.

Sono et al. [9] highlight the importance of adaptation for social and economic development, noting that effective adaptation measures are critical but often hindered by limited capacity and resources in many sub-Saharan African (SSA) countries.

Agricultural resilience strategies are essential for addressing the multiple challenges faced by farmers, particularly in sub-Saharan Africa, where climate change, natural disasters, and economic volatility increase vulnerability [46]. Crop diversification is a cost-effective method to enhance resilience by suppressing pest outbreaks, buffering against climate variability, and improving overall production stability [47].

Several strategies have been proposed to enhance resilience, including income and asset management, improved farm production techniques, government support programs, and technological development [46]. Adaptation measures, such as implementing improved crop varieties and efficient resource use, can address medium-term impacts [48]. The use of ICT-based services and modern technology also helps in coping with climate change [49].

Strengthening farmers' organizations, implementing gender-sensitive policies, and exploring innovative financing is crucial for enhancing resilience [49]. At the local level, strategies, such as selling livestock, alternative land use, and food storage, are common coping mechanisms [50]. Participation in social networks, access to resources, and diversified farming practices significantly contribute to household resilience against agricultural drought [50].

Cabo Verde's agriculture faces significant challenges, but farmers have developed various resilience strategies over time. These strategies include modern irrigation techniques for commercial farming and traditional rainfed subsistence systems that rely on social institutions for resource sharing. The government has implemented soil and water conservation measures, which have positively impacted both livelihoods and the environment [51].

To mitigate water scarcity, small-scale irrigation and drip technology have been introduced, enhancing agricultural production in arid areas [26]. Diversification of income sources, including remittances from emigrants, plays a crucial role in the resilience of rural families, particularly those with limited agricultural resources and female-headed households [52]. Despite these efforts, prolonged droughts, erosion, and soil degradation continue to present significant challenges to agrarian development across the archipelago (**Figure 3**) [26].

Agriculture is a strategic sector for Cabo Verde's gross domestic product, accounting for 8% of its composition. However, it faces significant challenges, including low technology adoption, limited land availability, soil fertility loss, and the advance of invasive plants, which contribute to a decline in agricultural practice. The land available for agriculture totals approximately 79,000 hectares or 19.6% of the total land area. The distribution of land use is as follows: arable land (12.9%), permanent crops

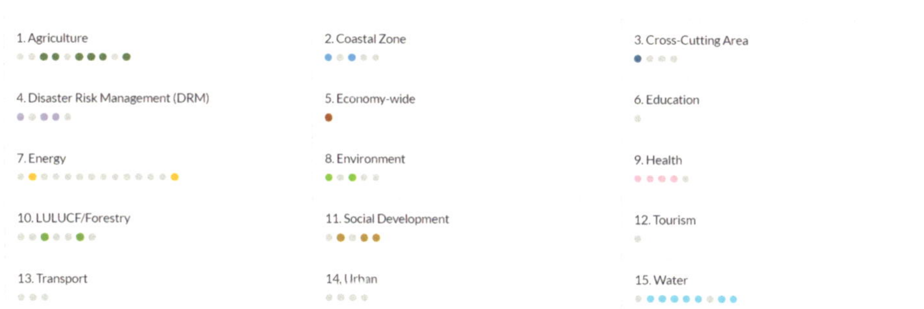

Figure 3.
Adaptation actions by sector (Agriculture) – Cabo Verde (The colored symbols represent adaptation actions prioritized by Cabo Verde in Updated First Nationally Determined Sector. The color gray symbols represent areas where there are currently no adaptation actions. These dots are mapped against critical systems that need an acceleration of adaptation efforts to address the impacts of climate change [53]). Source: Watch Data, [53].

(0.99%), permanent meadows and pastures (6.2%), forest area (22.07%) with an average annual reforestation rate of 4.14%, and other land uses (57.84%) [54].

The agricultural production system primarily relies on subsistence farming, focusing on crops, such as maize, which represents 44.2% of the total area, legumes at 43%, and vegetables at 4% of the total land use area [54]. Cabo Verde remains highly dependent on food imports, accounting for around 80% of its food supply [27], which significantly impacts family purchasing power, especially during international crises, such as COVID-19 and the Russia-Ukraine war, leading to increased prices for essential items. In response, the government intervened to stabilize prices, but the most vulnerable populations still experienced the effects of rising energy and food costs [55].

A priority for the agricultural sector is to reverse and prevent land degradation through soil and water conservation techniques, including mulching, planting cover crops, applying organic amendments, adopting agroforestry systems, utilizing efficient nondepleting biomass for energy, favoring nitrogen-fixing leguminous crops over soil-depleting ones, and implementing anti-erosion practices, such as terraces, contour ridges, and vegetation barriers.

To combat pests, the country is committed to promoting integrated pest and disease management, which includes using natural predators and approved pesticides to reduce seasonal infestations and crop yield losses [10].

In the water sector, a priority action is to improve the infiltration and replenishment of water resources through nature-based solutions (NbSs), such as soil cover for humidity retention, altitude moisture and rain harvesting, slope stabilization, and agroforestry [10].

To adapt agriculture to climate change and new technologies, investment is considered essential in agricultural research, extension, demonstration, and farmer capacity building. This includes modernizing traditional coping strategies for extreme weather and agronomic conditions, such as vegetative rain- and mist-harvesting, seed and water conservation, crop associations, and fruit tree cultivation. Additionally, developing locally appropriate, low-tech, low-energy, low-cost adaptation practices, and climate-adapting local plant and animal genetic resources and varieties—including short-cycle and drought-resistant crops—aims to enhance climate resilience and improve food quantity and nutritional quality [10].

6. Challenges and opportunities

Cabo Verde's vulnerability to climate change is exacerbated by socio-economic inequalities and widespread poverty. High levels of poverty and inequality further complicate the country's ability to manage disaster risks effectively. These factors limit access to resources and opportunities for vulnerable populations, making them more susceptible to the impacts of climate change. Rapid and unplanned urbanization also poses additional challenges, leading to increased vulnerability and complicating disaster risk management efforts [1].

The country faces additional risks, including landslides, erosion, deforestation, and forest fires [4].

Climate change adaptation (CCA) strategies in Cabo Verde include various approaches at different levels, from community to national. At the EU level, adaptation strategies often involve improvements in agricultural systems, such as crop diversification or the introduction of climate-resilient crop varieties [6]. However, in

Cabo Verde, there is minimal preservation of native seeds, with the matrix of existing seeds being lost due to the lack of transmission of knowledge and traditional practices to younger generations, as well as the abandonment of agricultural fields.

Risk assessments and associated plans aim to identify and mitigate potential risks faced by communities, including measures to safeguard ecosystems affected by climate change. Effective risk management requires continuous learning from inaction and mistakes. For the Cabo Verdean government, this process is ongoing and involves efforts in planning, risk screening, performance evaluation, and collaboration across communities and institutions [10].

Mechanisms to inform communities about imminent risks related to climatic events are in place; however, communities have not fully recognized that pests and irregular rainfall patterns are not solely due to historical rainfall variability and the islands' geographical situation but also due to climate change. There is a pressing need to invest in environmental education at all levels, from the community to the national level, including educational institutions, to promote understanding and preparedness for climate risks among community members.

The country also lacks sufficient management and conservation practices for water sources to ensure their availability in the face of climate change. Sustainable practices for protecting water sources, where water originates, must be implemented as many people perceive this resource as infinite.

Resilience to gender mainstreaming can enhance the effectiveness of climate change processes through resilient measures; conversely, exclusion threatens progress toward gender equality by silencing half of the global population, denying women their rights, and depriving society of their unique contributions, experiences, and skills crucial for poverty reduction and sustainable development [56].

There is a lack of adequate reporting on the impacts of disasters, particularly droughts and extreme temperatures. This gap hinders the understanding of the real consequences of these events and the effectiveness of disaster risk reduction strategies [1].

During the preparation of the post-disaster needs assessment (PDNA) document for the volcanic eruption in Fogo (2014–2015), proposals for disaster risk reduction were made. These included strengthening the early warning system for floods and droughts at the national level, improving technical capacity and equipment, enhancing responsiveness, and creating a national risk information system. This system is intended to bolster early warning, evaluation, and monitoring, serving as a tool for risk reduction, response, and recovery decisions [4].

The government has implemented soil and water conservation measures, such as terraces, dams, and afforestation, to address land degradation and water scarcity [51]. However, there has been a historical focus on reacting to disasters rather than implementing preventive measures. This reactive approach exacerbates the underlying factors driving disaster risk, particularly those related to climate change [1].

Costa [1] advocates for a comprehensive national framework that integrates disaster risk management with sustainable development goals. This involves consolidating achievements in risk reduction and climate resilience while integrating these efforts into broader poverty reduction strategies. Policymakers should prioritize developing local expertise and resources to reduce dependency on international aid, which can limit the adoption of tailored, sustainable solutions [1].

Engaging local communities in the planning and implementation of disaster risk management (DRM) strategies is vital. This participatory approach ensures that the needs and knowledge of those most affected by disasters are incorporated into decision-making processes, leading to more effective and sustainable outcomes [1].

A lack of coordination and duplication of tasks are highlighted as key issues in several public policy and consultancy documents related to climate change. This underscores the need to create a dedicated body to address climate change, serving as a bridge between responsible institutions. There is a high expectation that the new climate governance framework, by establishing a body for climate action, will address this shortcoming.

As a small island developing state (SIDS), Cabo Verde lacks sufficient financial resources to combat climate change without the support of developed countries and multilateral institutions [13]. This situation necessitates the development of climate diplomacy focused on negotiations and the mobilization of financial resources to implement adaptation plans. Cabo Verde needs to mobilize 2 billion euros over 10 years: 1 billion for adaptation and the remainder for mitigation measures.

Technologically, the country must overcome barriers to the diffusion of low-carbon technologies, involving private sector participation, and improving access to satellites, radars, and statistical and digital systems to generate efficient data for decision-making and action. Building resilience requires accurate climate data and scenarios. Cabo Verde needs to foster international partnerships to develop technologies and tools to produce strategic data.

7. Climate risk and uncertainty analysis

Cape Verde has a relatively small population, which presents challenges in achieving economies of scale in climate adaptation efforts, but it also allows for more targeted and potentially effective interventions with proper policies. A significant portion of the population is of working age, which could be leveraged for labor-intensive adaptation strategies, such as building resilient infrastructure or restoring ecosystems. Therefore, it is important to create socio-economic opportunities for the young population to contribute to the sustainable development of the country.

Population growth indicates increasing pressure on resources, especially in terms of water, food, and energy, all of which are vulnerable to climate change impacts. Cities face the challenge of greater population concentration, leading to increased pressure on these resources, while rural areas, which depend on agriculture, have faced droughts, pushing the population toward cities and emigration.

In the health sector, although the population is living longer, climate change exacerbates health risks, especially for vulnerable populations like the elderly. In this sense, events, such as droughts and floods, could disrupt healthcare services, particularly in rural areas.

Although Cape Verde's economy is relatively small, climate impacts on strategic sectors, including tourism, agriculture, and fisheries, lead to a disproportionate effect on GDP and economic growth projections due to the vulnerabilities of these sectors to climate change.

In the education sector, high internet penetration facilitates awareness, education, and communication regarding climate risks and adaptation strategies. Similarly, a high literacy rate enhances the population's ability to understand and engage with climate-related information, boosting resilience. Another important factor is the high enrollment rate and gender parity in education, which strengthens adaptive capacity by fostering a well-informed and empowered population capable of engaging in climate action.

Institutions play an important role in tackling climate change. In this sense, the moderate level of trust in the police and members of parliament hinders the enforcement of climate-related regulations and the implementation of necessary policy changes, especially when public concerns about corruption are added. These concerns undermine climate adaptation efforts as funds meant for climate projects might be misallocated or misused (**Table 2**).

7.1 Climate future

The projection considers different emission scenarios or *shared socio-economic pathways (SSPs)*. The model offers five paths that countries can follow; therefore, it offers alternatives to plausible futures and considers a series of social, economic,

Topics	Indicators	Values	Year Ref.
Population	Total Population	506.595	2023
	15–64 years	67.18%	
	Population growth	1.19%	2023
Health	Life expectancy at birth (years)	74	2021
	Immunization, measles (% of children aged 12 to 23 months)	95	2021
	Proportion of maternal mortality (per 100,000 live births)	42	2020
	Infant mortality rate/1000 live births	22.96	2023
	Density of doctors/1000 inhabitants	0.83	2018
Economy	Gross Domestic Product (US$) billion	2.29	2022
	GDP per capita (USD)	3902.6	2022
	Economic growth forecast	5.3%	2025
	Youth unemployment rate	34.3%	2021
Education	Individuals using the Internet (% of population)	70	2021
	School enrollment, primary (% gross)	101	2019
	School, primary and secondary enrollment (gross), gender parity index (GPI)	1.01 [17]	2019
	Literacy rate, total adults (% of people aged 15 and over)	91	2021
	Proportion of seats held by women in national parliaments (72)	39%	2022
Trust in institutions:	Public education system	85.7% trust	2019
	Public health system	69.0% trust	
	Police	59.4% trust	
	Members	42.9% trust	
	Level of population concern about corruption	75,2% concern	

Source: WFP [57]; WHO, 2024, World Bank [58]; INE-CV [59, 60].

Table 2.
Key indicators.

cultural, technological, institutional, and biophysical characteristics, which characterize the interactions between human and natural systems and outline visions for the future, on a specific scale (**Figure 4**) ([61], p. 555).

Scenario narratives are qualitative descriptions of plausible future world evolutions (applied to countries) based on a coherent and consistent set of assumptions about key driving forces, such as technological change [61]. These scenarios present baselines of how things would look in the absence of climate policy, allowing researchers to examine barriers and opportunities for climate mitigation and adaptation in each possible future world when combined with mitigation targets [62].

Scenarios 1 and 5 are more optimistic for human development if investments in education and health are made, alongside the proper functioning of institutions. However, while Scenario 5 envisions this development being driven by a fossil fuel-based economy, Scenario 1 assumes it will be fueled by a growing shift toward sustainable practices, ensuring more inclusive development that complies with environmental limits [62].

In contrast, SSP3 and SSP4 are more pessimistic scenarios for economic and social development due to low investment in education and health in poorer countries, which will face rapidly growing populations. Finally, SSP2 represents a "middle-of-the-road" scenario, mirroring the historical development patterns followed throughout the twenty-first century [62].

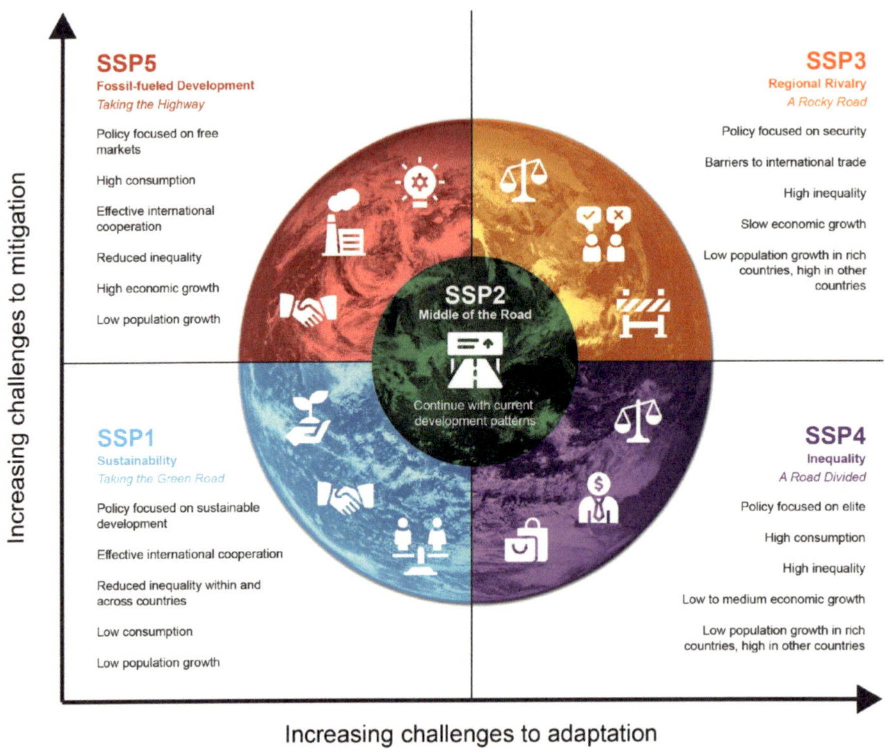

Figure 4.
Shared socio-economic pathways (SSP). Source: climate data (Understanding Shared Socio-economic Pathways (SSPs) — ClimateData.ca), 2024.

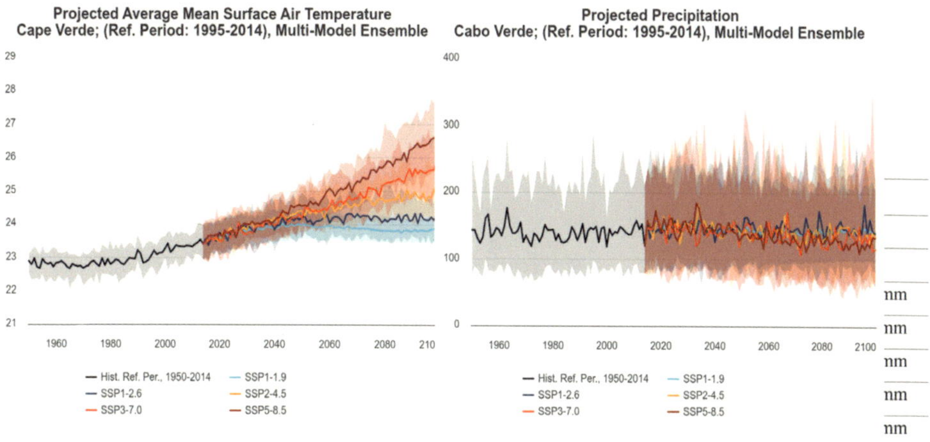

Figure 5.
Temperature and precipitation projection (2040–2100). Source: World Bank Climate Change Knowledge Portal [25].

Considering the scenarios, Cape Verde presents the following data related to temperature and precipitation (**Figure 5**).

7.1.1 Temperature

SSP1–1.9 and SSP1–2.6 (the most optimistic scenarios with stringent climate policies): Temperatures remain relatively stable, with slight increases over time. The rise is modest, with temperatures peaking around 24.15°C in SSP1–2.6 by 2080. These scenarios suggest effective climate mitigation efforts, leading to controlled warming (**Table 3**).

SSP2–4.5 (a "middle-of-the-road" scenario): Temperatures rise more significantly, reaching 24.90°C by 2100. This suggests a moderate level of warming, with less effective climate policies than SSP1 scenarios.

SSP3–7.0 and SSP5–8.5 (the most pessimistic scenarios with high emissions): These scenarios predict the highest temperature increases. By 2100, SSP3–7.0 reaches 25.82°C, and SSP5–8.5 reaches 26.58°C. The increase in temperature is particularly sharp after 2060, indicating a significant rise in global warming without effective climate policies.

7.1.2 Precipitation

As for the precipitation projections, the SSP1–1.9 and SSP1–2.6 levels fluctuate slightly but remain relatively stable, with values ranging from 126.46 to 151.60 mm. These scenarios indicate that, despite some variability, the impacts on precipitation may be manageable under strong climate policies.

In SSP2–4.5 precipitation shows more variation, peaking at 151.80 mm in 2060, then dropping to 126.77 mm in 2080, and stabilizing again by 2100. This suggests more unpredictable rainfall patterns compared to the SSP1 scenarios.

The SSP3–7.0 and SSP5–8.5 show significant declines in precipitation over time in Cabo Verde, especially in SSP3–7.0, where precipitation drops from 156.82 mm in 2060 to just 112.70 mm by 2100. SSP5–8.5 also sees a decline in precipitation, although less dramatic. These trends indicate potential challenges for water resources and agriculture under higher-emission scenarios.

Scenarios	Temperature				Precipitation			
	Period				Period			
	2040	2060	2080	2100	2040	2060	2080	2100
SSP1–1.9	23.86°C	23.94°C	23.81°C	23.88°C	139.06 mm	146.94 mm	141.18 mm	141.60 mm
SSP1–2.6	23.84°C	24.04°C	24.15 °C	24.13°C	126.46 mm	146.56mm	151.60 mm	142.26 mm
SSP2–4.5	24.21 °C	24.59°C	24.68°C	24.90°C	140.77 mm	151.80 mm	126.77 mm	141.11 mm
SSP3–7.0	23.97°C	24.56°C	25.22°C	25.82°C	140.70 mm	156.82 mm	116.26 mm	112.70 mm
SSP5–8.5	24.10 °C	25.08°C	25.77 °C	26.58°C	143.06 mm	138.75 mm	124.56 mm	133.26 mm

Source: World Bank Climate Change Knowledge Portal [25].

Table 3.
Temperature and precipitation values for the selected period.

Across all scenarios, temperatures rise, with the most dramatic increases in SSP3–7.0 and SSP5–8.5. These increases are likely to exacerbate heat-related challenges, including stress on ecosystems, human health, and infrastructure. The scenarios indicate increased variability in precipitation, with potential declines in rainfall in later years, especially in high-emission scenarios. This could lead to challenges in water availability and agricultural productivity.

The data underscore the importance of strong climate policies (as seen in SSP1 scenarios) in limiting temperature increases and stabilizing precipitation patterns. In contrast, the higher-emission scenarios (SSP3–7.0 and SSP5–8.5) suggest more severe impacts, highlighting the risks of inaction.

7.1.3 Variation by season and region (islands)

Regarding seasonal variation, the hottest season in Cabo Verde generally includes June to August, with temperatures projected to reach as high as 29.79°C (SSP3–7.0) in 2080–2099. While the coolest season covers December to February, temperatures are still projected to rise. Under SSP3–7.0, the temperature rises from 26.61°C in 2020–2039 to 27.29°C by 2080–2099.

The regional variation points to islands, such as Maio and Boa Vista, which are projected to experience the highest temperatures. For example, under SSP3–7.0, Boa Vista's temperature during JJA increases from 28.93°C (2020–2039) to 30.58°C (2080–2099). Meanwhile, Santo Antão consistently shows the lowest temperatures across the projections but still exhibits significant warming. For instance, its DJF temperature under SSP1–1.9 rises from 25.54°C (2020–2039) to 25.61°C (2080–2099).

These variations imply that increasing temperatures, especially during the summer months, may exacerbate heat stress for both the population and ecosystems, affecting two vital sectors: water resources and agriculture. Higher temperatures lead to increased evaporation rates, potentially exacerbating water scarcity and affecting agricultural productivity, particularly for temperature-sensitive crops.

7.2 Risk and uncertainty analysis

Risk is the combination of the probability of an event and its negative consequences, resulting from the specific impact of a hazard on preexisting conditions of vulnerability [4]. In this context, risk management deals with uncertainties to minimize potential damage and losses. Uncertainty exists for any future projection, in this regard, the rate of future global warming depends on future emissions, feedback processes that dampen or reinforce disturbances to the climate system, and unpredictable natural influences on climate, such as volcanic [58].

The Scenarios - SSP1 promises a more optimistic future compared to the other scenarios, but presupposes robust investments in education and health, in alliance with the efficient use of rapid economic growth, as well as the excellent functioning of institutions [62].

In this scenario, Cape Verde aims to transition toward a more sustainable economic model, emphasizing the importance of integrating environmental considerations into its development strategies. The government's strategic sustainable development plan (PEDS I and II) outlines goals for economic diversification and resilience against climate impacts, aligning with the sustainability objectives of SSP1. However, Cape Verde is classified among the most economically vulnerable countries in Africa, with a heavy reliance on tourism as a primary growth sector. This dependence creates risks, especially under scenarios that involve economic shocks or climate impacts that could deter tourism, especially in the main tourist destinations, such as Sal, Boa Vista, and Maio, which may face rising sea levels.

The SSP2 reflects a continuation of current trends with moderate challenges, Cape Verde's reliance on tourism and external markets places it in a fragile position, where socio-economic vulnerabilities are pronounced. The country still faces significant challenges, such as those experienced during the COVID-19 pandemic, which tested the country's responsiveness, while still demonstrating that it is a country dependent on foreign aid. In addition, the Russia-Ukraine war has interfered with the international prices of food and fuel products, affecting the purchasing power of Cape Verdean families, compromising their level of savings and income.

Under the SSP3 scenario, global cooperation diminishes, leading to heightened competition for resources. Cape Verde could face huge challenges, such as food insecurity and increased migration pressures due to climate impacts. These challenges are already beginning to manifest, with an agricultural deficit and scarcity of rain on the main agricultural islands, such as Santo Antão, Santiago, and Fogo, affecting production and family income. A relevant fact to be added is the irregularity of interisland transport, which affects the shipment of products to the main national markets.

The SSP4 scenarios indicate the country faces social inequalities that could be intensified due to the concentration of economic activity in tourism where the urban areas lead to increased disparities between urban and rural populations. The all-inclusive tourism model, while recruiting a considerable number of local employees, working conditions and contracts are precarious, and it fails to encompass the small economic activities around it, which contribute to the empowerment of local communities.

SSP5 emphasizes, although Cape Verde has made strides toward renewable energy, reliance on fossil fuels remains a concern. This scenario emphasizes rapid economic growth driven by fossil fuel consumption, which could conflict with Cape Verde's sustainability goals and its need to adapt to climate change.

8. Final remarks

Climate change presents a unique opportunity for Cabo Verde to invest in both economic and climate prosperity. The state must play a decisive role in this ecological transition by formulating policies, creating lines of credit, and financing initiatives that support sustainable domestic development. Leveraging traditional knowledge and practices in agriculture, livestock, fisheries, and forestry can help build a national repository of nature-based technologies, enhancing the country's response to climate impacts.

For resilience to be achieved, climate considerations must permeate all public policy agendas. Ideally, climate change should be integrated into every aspect of national policy, ensuring that it is central to sectoral agendas. This approach would facilitate greater flexibility and quicker policy innovation, aligning with the global emphasis on climate issues and the urgent need for action.

An environmental problem is inherently linked to social and economic inequalities. Thus, reconfiguring the national agenda from a climate change perspective offers a strategic opportunity to address deeper structural and complex issues. This approach ensures that critical sectors, such as agriculture, receive the attention they need.

The concept of resilience requires internal mechanisms capable of restoring systems to normalcy. Vulnerable countries, such as Cabo Verde, must build their own capacities to manage and mitigate damage and losses, rather than relying solely on external support. Strengthening these internal mechanisms is crucial so that, in the face of extreme climatic or natural events, the country can rely on its own resources and capabilities.

To ensure effective functioning, all components of the climate agenda must be interconnected. This analogy underscores the need to identify and address any gaps in the climate strategy and internal capacities. Gender considerations, particularly in the agriculture sector, require greater emphasis. Women should not only be recognized as victims of climate change but also as transformative agents who can contribute to solutions, provided they are granted equal socio-economic opportunities.

The variation in temperature across regions (islands) presents two key insights across the scenarios that must be considered. The islands that will be most affected are the tourist islands: Sal, Maio, and Boa Vista. On the other hand, the temperature indicates that the sector most impacted will be agriculture. These are two strategic sectors for Cabo Verde, which must make efforts to transform tourism into a more sustainable sector and agriculture into one that is resilient to climate change.

Author details

Pedro Andrade Matos
Departamento de Direito, Universidade de Santiago (US), Assomada, Cabo Verde

*Address all correspondence to: matooscv@hotmail.com

References

[1] Ferreira Costa CG. Comprendiendo y reduciendo los riesgos climáticos: El impacto de políticas innovadoras para una respuesta sostenible a las sequías en Cabo Verde. Geographical Studies. 2020;**81**(288):e033

[2] Intergovernmental Panel on Climate Change (IPCC), Metz B. Climate Change 2007: Mitigation of Climate Change. Summary for Policymakers. In: Contribution of Working Group III to the Fourth Assessment Report of the Intergovernmental Panel on Climate Change. Cambridge: Cambridge University Press; 2007. Available from: https://www.ipcc.ch/report/ar4/wg3/

[3] Siebert C. Pós-desastre: reconstruindo a armadilha. Jornal Expressão Universitária. 1 Feb 2011:6-7

[4] Verde C. Avaliação das necessidades pós-desastre (PDNA): Erupção vulcânica no Fogo 2014-2015. Praia: GRF; 2015

[5] World Bank. Climate and Disaster Resilience Financing in Small Island Developing States. Paris: OECD and International Bank for Reconstruction and Development/The World Bank; 2016

[6] Mercer J. Disaster risk reduction or climate change adaptation: Are we reinventing the wheel? Journal of International Development. 2010;**22**(2):247-264. DOI: 10.1002/jid.1677

[7] World Bank. Building resilience: Integrating climate and disaster risk into development. In: Lessons from World Bank Group Experience. New York: The World Bank; 2013. Available from: http://www.worldbank.org/content/dam/Worldbank/document/SDN/Full_Report_Building_Resilience_Integrating_Climate_Disaster_Risk_Development.pdf

[8] IPCC. In: Barros VR, Field CB, Dokken DJ, Mastrandrea MD, Mach KJ, Bilir TE, editors. Climate Change 2014: Impacts, Adaptation, and Vulnerability. Part B: Regional Aspects. Contribution of Working Group II to the Fifth Assessment Report of the Intergovernmental Panel on Climate Change. Cambridge: Cambridge University Press; 2014

[9] Sono D, Wei Y, Jin Y. Assessing the climate resilience of sub-Saharan Africa (SSA): A metric-based approach. Land. 2021;**10**:1205. DOI: 10.3390/land10111205

[10] Verde C. Plano Nacional de Adaptação de Cabo Verde. República de Cabo Verde: Direção Nacional do Ambiente, Praia; 2021

[11] Eing C. Capacidades estatais em adaptação às mudanças climáticas na América Latina. Santiago: Latino Adapta: Policy Brief; 2019

[12] Meckling J, Nahm J. The power of process: State capacity and climate policy. Governance. 2018;**31**:1-17. DOI: 10.1111/gove.12338

[13] Thomas A et al. Climate change and small island developing states. Annual Review of Environment and Resources. 2020;**45**:1-27

[14] Klöck C, Nunn PD. Adaptation to climate change in small island developing states: A systematic literature review of academic research. The Journal of Environment & Development. 2019;**28**(2):196-218. DOI: 10.1177/1070496519835895

[15] Kelman I, West JJ. Climate change and small island developing states: A critical review. Ecological and Environmental Anthropology. 2009;**5**:1-16

[16] Robinson S. Adapting to climate change at the national level in Caribbean small island developing states. Island Studies Journal. 2018;**13**(1):79-100. DOI: 10.24043/isj.59

[17] Bahers J-B, Singh S, Durand M. Analyzing socio-metabolic vulnerability: Evidence from the Comoros archipelago. Anthropocene Science. 2022;**1**(1):164-178. DOI: 10.1007/S44177-022-00017-1. Available from: https://hal.science/hal-03685485

[18] Moncada S, Bambrick H, Briguglio L, Iorns C, Kelman I, Nurse L. Introduction to the book. In: Moncada S, Briguglio L, Bambrick H, Kelman I, Iorns C, Nurse L, editors. Small Island Developing States. Vol. 9. Malta: Springer; 2021. DOI: 10.1007/978-3-030-82774-8_1

[19] Varela D, Romeiras MM, Silva L. Implications of climate change on the distribution and conservation of Cabo Verde endemic trees. Global Ecology and Conservation. 2022;**34**:e02025

[20] Lacerda F, Gmb L, Assad ED. Long-term temperature and rainfall trends over Northeast Brazil and Cabo Verde. Journal of Earth Science & Climatic Change. 2015;**6**:1-8

[21] Tavares J, Lopes M, Lima-Neto FB. Climate and fundamentals of the energy offer in Cabo Verde. Energy Reports. 2020;**6**:370-377

[22] Matos PA, Andrea G, Garcia F, Aparecida M, Santos D. The role of gender in mitigation and adaptation to climate change in Cabo Verde. Veredas do Direito. 2023;**20**:e202536. Direito Ambiental e Desenvolvimento Sustentável

[23] MrGeogWagg. Climate Change and the Sahel Region. 2015. Available from: https://mrgeogwagg.wordpress.com/2015/10/05/climate-change-and-the-sahel-region/ [Accessed: October 10, 2024]

[24] Neves NA. Género e alterações climáticas: A influência de gender mainstreaming na implementação de projetos de desenvolvimento [Master's thesis]. Instituto Superior de Economia e Gestão; 2019. Available from: https://www.repository.utl.pt/handle/10400.5/19138

[25] World Bank Climate Change Knowledge Portal. Cape Verde: Current Climate Climatology. 2024. Available from: https://climateknowledgeportal.worldbank.org/country/cape-verde/climate-data-historical [Accessed: October 10, 2024]

[26] Monteiro F, Fortes A, Ferreira V, Pereira Essoh A, Gomes I, Correia AM, et al. Current status and trends in Cabo Verde agriculture. Agronomy. 2020;**10**(1):74

[27] Mendoza-Grimón V, Fernández-Vera JR, Silva GD, Semedo-Varela A, Palacios-Díaz MD. Cabo Verde (West Africa) successful water reuse pilot project: A sustainable way for increasing food production in a climate change scenario. Water. 2021;**13**(2):160

[28] Semedo J, Borges A. Os Cabo-verdianos defendem a igualdade de direitos entre homens e mulheres mas admitem que o governo precisa fazer mais. Afrobarometer Edição No. 730; 2023

[29] Eduonoo M. Climate Change Worsens Life in Cabo Verde; Citizens

Want Collective Action to Combat it. Accra: Afrobarometer Dispatch, No. 726; 2023

[30] Food and Agriculture Organization of the United Nations. Cabo Verde. 2024. Available from: https://www.fao.org/faostat/en/#country/35

[31] Instituto Nacional de Estatística - INE. Estatísticas das condições de vida dos agregados familiares: Inquérito Multiobjetivo Contínuo 2023. Praia: INE; 2024

[32] Sultana F. Gender and Water in a Changing Climate: Challenges and Opportunities. Cham: Springer International Publishing; 2018

[33] Huyer S. Gender and International Climate Policy: An Analysis of Progress in Gender Equality at COP21. Montpellier: CCAFS Info Note; 2016

[34] Pandit A, Batelaan O, Panta SK, Pandey VP, Adhikari S. The gendered implication of declining spring sources in the Rangun watershed area. Far Western Review. 2023;**1**(2):101-116

[35] Shrestha S, Chapagain PS, Ghimire M. Gender Perspective on Water Use and Management in the Context of Climate Change: A Case Study of Melamchi Watershed Area, Nepal. Thousand Oaks: SAGE Open; 2019. p. 9

[36] The World Bank Group, Asian Development Bank. Climate Risk Country Profile: Timor-Leste. 2021

[37] Al Khourdajie A, van Diemen R, Lamb WS, Pathak M, Reisinger A, Skea JE, et al. IPCC, 2022: Annex II: Definitions, Units and Conventions. 2022

[38] Cartier R, Barcellos C, Hübner C, Porto MF. Vulnerabilidade social e risco ambiental: uma abordagem metodológica para avaliação de injustiça ambiental. Cadernos de Saúde Pública. Dec 2009;**25**(12):2695-2704

[39] Khan NA, Gao Q, Abid M. Public institutions' capacities regarding climate change adaptation and risk management support in agriculture: The case of Punjab Province, Pakistan. Scientific Reports. 2020;**10**:1-12

[40] Islam MT, Nursey-Bray M. Adaptation to climate change in agriculture in Bangladesh: The role of formal institutions. Journal of Environmental Management. 2017;**200**:347-358

[41] Meinzen-Dick RS, Bernier Q, Haglund E. The Six "Ins" of Climate-Smart Agriculture: Inclusive Institutions for Information, Innovation, Investment, and Insurance. Washington: International Food Policy Research Institute (IFPRI); 2013

[42] Shaw EE, Witt GB. Climate change and adaptive capacity in the Western Australian rangelands: A review of current institutional responses. Rangeland Journal. 2015;**37**:331-344

[43] Birk T, Rasmussen K. Migration from atolls as climate change adaptation: Current practices, barriers and options in Solomon Islands. Natural Resources Forum. 2014;**38**(1):1-13. DOI: 10.1111/1477-8947.12038

[44] Campbell D, Barker D, McGregor D. Dealing with drought: Small farmers and environmental hazards in southern St Elizabeth, Jamaica. Applied Geography. 2011;**31**(1):146-158. DOI: 10.1016/j.apgeog.2010.03.007

[45] Chandra A, Dargusch P, McNamara KE. How might adaptation to climate change by smallholder farming

communities contribute to climate change mitigation outcomes? A case study from Timor-Leste, Southeast Asia. Sustainability Science. 2016;**11**(3):477-492. DOI: 10.1007/s11625-016-0361-9

[46] Chuku C, Okoye C. Increasing resilience and reducing vulnerability in sub-Saharan African agriculture: Strategies for risk coping and management. African Journal of Agricultural Research. 2009;**4**:1524-1535

[47] Lin BB. Resilience in agriculture through crop diversification: Adaptive management for environmental change. BioScience. Mar 2011;**61**(3):183-193. DOI: 10.1525/bio.2011.61.3.4

[48] Speranza IC. Resilient adaptation to climate change in African agriculture. In: Studies/Deutsches Institut für Entwicklungspolitik: Vol. 54. Bonn: Deutsches Institut für Entwicklungspolitik (DIE)/German Development Institute (GDI); 2010

[49] Shimeles A, Verdier-Chouchane A, Boly A. Conclusions: Enhancing the resilience and sustainability of the agriculture sector in sub-Saharan Africa. In: Shimeles A, Verdier-Chouchane A, Boly A, editors. Building a Resilient and Sustainable Agriculture in Sub-Saharan Africa. Cham: Palgrave Macmillan; 2018

[50] Bahta YT, Myeki VA. Adaptation, coping strategies and resilience of agricultural drought in South Africa: Implication for the sustainability of livestock sector. Heliyon. 2021:7

[51] Baptista I, Fleskens L, Ritsema CJ, Querido A, Tavares JD, Ferreira AJ, et al. Soil and water conservation strategies in Cabo Verde (Cabo Verde in Portuguese) and their impacts on livelihoods: An overview from the Ribeira Seca watershed. Land. 2015;**4**:22-44

[52] Reis VR. Fontes de rendimento das famílias rurais e a segurança alimentar em Cabo Verde. Cadernos de Estudos Africanos. 2015;**29**:129-157

[53] Climate Watch Data. Cabo Verde. 2023. Available from: https://www.climatewatchdata.org/countries/CPV?end_year=2021&start_year=1990 [Accessed: October 10, 2024]

[54] Food and Agriculture Organization of the United Nations. FAOSTAT Cabo Verde. 2019. Available from: https://www.fao.org/faostat/en/#country/35

[55] Expresso das Ilhas. Efeito da guerra na Ucrânia em Cabo Verde é mais gravoso do que a pandemia – PM. Praia: Expresso das Ilhas; 2022. Available from: https://expressodasilhas.cv/economia/2022/06/30/efeito-da-guerra-na-ucrania-em-cabo-verde-e-mais-gravoso-do-que-a-pandemia-pm/80832

[56] Hemmati M, Röhr U. Engendering the climate-change negotiations: Experiences, challenges, and steps forward. Gender and Development. 2009;**17**(1):19-32. Available from: https://www.tandfonline.com/doi/abs/10.1080/13552070802696870?tab=permissions&scroll=top

[57] World Food Programme. Annual Country Report 2023: Cape Verde. 2023. Available from: https://www.wfp.org/operations/annual-country-report?operation_id=CV01&year=2023#/26108 [Accessed: October 10, 2024]

[58] World Bank Group. Cape Verde High-level Summary: Compound Heat Risk. 2024. Available from: https://climateknowledgeportal.worldbank.org/country/cape-verde/heat-risk [Accessed: October 10, 2024]

[59] Instituto Nacional de Estatística. Estatísticas das condições de vida

dos agregados familiares: Inquérito Multiobjetivo Contínuo 2022. Praia: INE; 2022

[60] Instituto Nacional de Estatística. Inquérito Multiobjectivo Contínuo 2016: Estatísticas de Governança, Paz e Segurança. Praia: INE; 2018

[61] Intergovernmental Panel on Climate Change (IPCC). Annex I: glossary. In: Global Warming of 1.5°C: An IPCC Special Report on the Impacts of Global Warming of 1.5°C Above Pre-Industrial Levels and Related Global Greenhouse Gas Emission Pathways, in the Context of Strengthening the Global Response to the Threat of Climate Change, Sustainable Development, and Efforts to Eradicate Poverty. Cambridge: Cambridge University Press; 2018. Available from: https://www.ipcc.ch/sr15/chapter/glossary/

[62] Hausfather Z. Explainer: How 'Shared Socioeconomic Pathways' Explore Future Climate Change. 2018. Available from: https://www.carbonbrief.org/explainer-how-shared-socioeconomic-pathways-explore-future-climate-change/ [Accessed: October 10, 2024]

Chapter 7

Perspective Chapter: One Step Forward, Two Steps Back – A Multilevel and Multi-Actor Analysis of the Importance of the Climate Emergency for Social Change in Brazil

Leila da Costa Ferreira, Fabiana Barbi, Niklas Weins and Luciana Lima Domingues de Souza

Abstract

This article analyzes Brazilian climate change policies from a multilevel and multi-actor perspective, seeking to understand the roles of the different actors in this process, both at the national and local levels. Since human activities are largely responsible for exacerbating global environmental change, especially climate change, understanding the social and political dimensions of these changes is essential in order to undertake effective strategies against the resulting impacts, emphasizing the issues of adaptation and mitigation. In this chapter, we map the emergence of specific municipal policies, institutional mechanisms, and their proposed mitigation and adaptation strategies. We found that the inclusion of different interested parties from society at different levels has contributed significantly to the resilience of advancing the climate change agenda in Brazil, as there have been major setbacks in the last decade due to the increasingly politically polarized environment. We propose public policies for urban development in favor of climate adaptation and mitigation and a call for inclusive and open policies that engage the population. We also recommend strengthening land use and occupation in Brazil, effectively coordinated with national and international climate governance.

Keywords: climate change, adaptation policies, multilevel analysis, climate citizenship, Brazil

1. Introduction

One of the latest Intergovernmental Panel on Climate Change (IPCC) communications reinforced that recent changes in climate are widespread, rapid, intensified, and unprecedented in at least 6500 years [1], with urban centers being the places most subject to their impacts. The climate emergency is already affecting various regions of the Earth, through rising sea levels, increasingly long periods of drought and urban heat islands, excessive precipitation, and consequent flooding as well as land displacement. It is now scientific consensus that all of these changes are undoubtedly caused by human activities.

Risks are the problems faced by societies based on the way they are organized, structured, and developed. In this way, it is safe to say that climate-type risks produce effects of multi-scalar and multidimensional processes. As climate change has become an ever more central social and political concern, sociologists like Anthony Giddens and Ulrich Beck have suggested that societal logic is fundamentally changing from the distribution of the "goods" of modernity to the management of the "bads." Despite an initially negative outlook on this emergence of risks like climate change, Beck suggested in his work on the "metamorphosis of the world" that climate change may actually turn into a force for good, putting into motion forces for unprecedented changes [2]. Thus, the sociological and analytical question that we are working on here is: in times of crisis, what does climate change do for society and how does it alter the order of Brazilian society and politics?

Whenever human activities are the most responsible for the worsening of global environmental changes, especially climate change, understanding the social and political dimensions of these changes is essential to undertake strategies to fight the impacts resulting from them [3]. As a consequence, the black box of traditional political issues reopens, inducing the need to overcome conventional social and political practices to forms of transnational responsibility in response to a global challenge. The causal mechanisms that generate the risks of climate change and the conditions for tackling them in times of globalized information society can now be clarified and identified [4].

The cosmopolitan metamorphosis of climate change (or global risk in general) concerns the coproduction of risk perceptions and normative horizons [2]. The idea of metamorphosis [2] is a new way of generating and implementing norms in the era of climate change. In the case of climate change as a metamorphosis, there is an agglutination between the topics of nature, society, and politics [5].

Climate change produces a basic sense of ethical and existential violation that could create new worldviews, norms, laws, markets, technologies, understandings of nations and states, urban forms, and international cooperation. Therefore, attention should be focused on what is emerging now like future structures and norms and new beginnings.

Addressing such a multifaceted challenge as the climate emergency, solutions are also expected to be comprehensive, including different areas of human activity, interested parties, and sectors of society, such as multilateral agencies, governments, the private sector, research institutes, and organized civil society groups. Furthermore, considering this is an anthropogenic challenge, characterized by its multidimensionality and its complex nature, it would be naive to believe that only one group of actors would be able to solve the climate crisis [6].

The involvement of nongovernmental actors, civil society organizations, private initiatives, universities, and research institutions are essential for the production of efficient and successful responses to the problem. These actors gain new political protagonism due to the centrality of climate change, forming multi-actor coalitions that reflect societal priorities and often anticipating governmental policies. These actor coalitions help to articulate knowledge, experience, and practices as climate change presents itself as a multiscale challenge, simultaneously relating to local and global scales.

There are several important actors in this process, and, among them, global cities and their networks are emerging as cosmopolitan actors. National governments continue to be relevant actors leading and coordinating the fight against climate change, in official and public mitigation and adaptation strategies. However, over the past decade, subnational levels (in particular cities) have gained political protagonism unseen before and have led forceful responses to climate change challenges around the world [7].

With regard to the multi-scalar and multidimensional aspects of the climate challenge, firstly, it is important to identify the main "players" in this conjuncture. Brazil is an important player, as it is among the top 10 cumulative emitters of greenhouse gases (GHG) (**Figure 1**). However, when observing its emissions profile, it becomes clear that its share of global fossil emissions is small, especially when considering cumulative emissions. It also becomes clear that most of its GHG emissions are due to changes in land use, a profile only similar to other big agricultural exporters like Indonesia and Argentina.

Land use and occupation, energy, and agriculture are emblematic sectors of the Brazilian emissions profile and are therefore largely linked to changes in land use and land management practices [9]. Based on the Brazilian Climate Change Observatory's data in the Greenhouse Gas Emissions and Removals Estimation System (SEEG) [9], in 2018, land use and land cover change contributed to 44% of the country's total emissions, followed by agriculture, which accounted for 25%. Deforestation has been the main source of land use emissions, accounting for 93% of the sector's total from 1990 to 2018. These are mostly driven by global demand for Brazil's agricultural commodities [10].

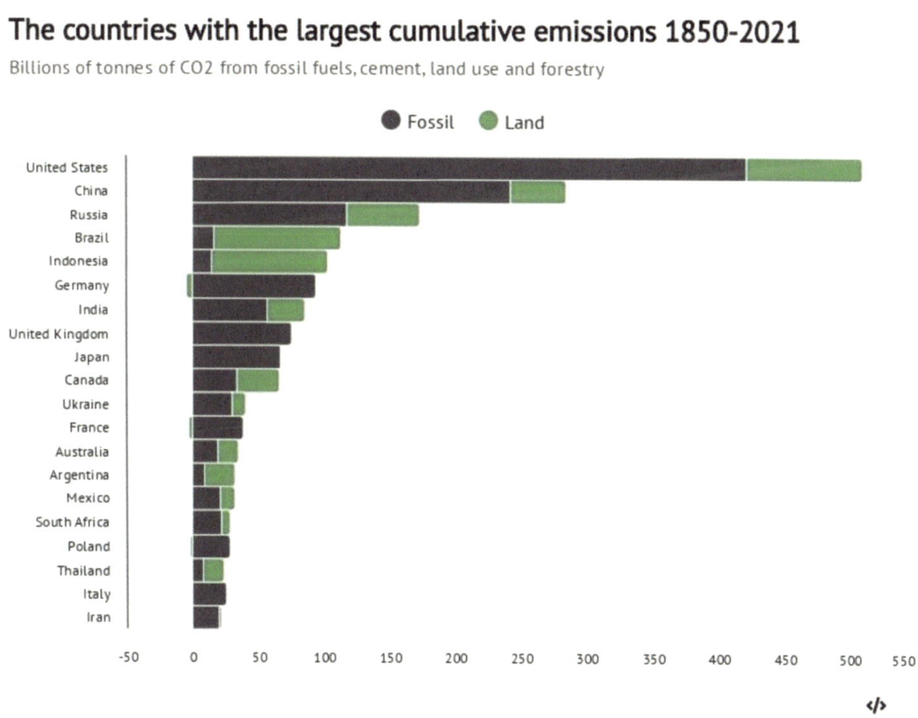

Figure 1.
Countries with the highest cumulative GHG emissions (1850–2021). Source: Carbon Brief [8].

However, a domestic factor is the advance of urbanization over green areas, a driver that is also an increasing part of the deforestation problem. Therefore, reducing deforestation, promoting reform of agricultural structures and land, and developing climate-oriented urbanism are key actions for an advanced climate agenda in the country. Brazil's often precariously urbanized cities concentrate on populous and poor peripheral zones, which are fragile to sudden and extreme climate changes. In terms of vulnerability to climate change impacts and extreme events, Brazil ranks 86 out of 181 countries in the ND-GAIN 2023 Index [11]. Extreme temperatures, rising sea levels, as well as the complex challenges of different regions across the country facing significant water scarcity and heavy rainfall are predicted to put significant pressure on vulnerable groups, urban infrastructure, the economy, and the country's unique ecosystems [12].

In the face of the climate emergency, rethinking subnational and local planning in the social, economic, and ecological spheres is an urgent priority [13]. Given this panorama, social actors play an essential role in the planning, articulation, and advocacy of public policies that promote mitigating practices and climate adaptation for the population [4–6]. Ultimately it is at the municipal level that governance is closest to the population, private initiative, infrastructure, and local public services [6, 13].

This chapter aims to analyze Brazilian climate change policies from a multilevel and multi-actor perspective, seeking to understand the roles at both the national and local levels, as well as the involvement of actors in this process.

2. Local dimension of climate change in Brazil

According to the World Economic Forum [14], almost half (44%) of global GDP from cities is at risk due to losses of nature and biodiversity. The report exposes failures in climate action that can impact the economies of municipalities around the world. In this case, the report emphasizes the positive role of biodiversity for local economies by influencing air quality, water cycles, and flood regulation, as well as supporting the production of energy, food, and medicines.

This highlights how the climate crisis is not just a problem to be resolved by the scientific community. It is also the most significant political and intellectual challenge of this era. Furthermore, it is inextricably linked to three other growing systemic crises, such as biodiversity loss, industrial pollution, and inequalities in economic, social, racial, and gender spheres. The climate, biodiversity, pollution, and inequality crises amplify each other and represent a larger systemic crisis affecting democracy, the neoliberalist system, and ultimately, civilization.

In Brazil, 87% of the population lives in urban areas, corresponding to 185 million people, undergoing an urban deficit arising from the urbanization process and urban problems intensifying with climate change, such as housing deficit, lack of basic sanitation, problems with water supply, among others [15]. However, our research has shown that Brazilian local climate policies are still relatively isolated initiatives in the national context and that there are few substantive advances.

By 2024, we identified 12 of the 5570 municipalities that have a specific law that establishes a climate policy, corresponding to a population of 31,126,829 [16]. Between 2003 and 2011, seven municipalities approved their public climate change policies (Belo Horizonte, Curitiba, Feira de Santana, Manaus, Palmas,

Rio de Janeiro, and São Paulo) and five municipalities approved them after 2014, with greater attention to adaptation (Fortaleza, Porto Alegre, Recife, Santos, and Sorocaba).

Not all of these municipal policies have clear mitigation or adaptation strategies. Seven of the 12 municipalities have mitigation strategies and six of them have adaptation actions [16]. Three municipalities did not define mitigation or adaptation actions. Most of the mitigation strategies include the establishment or planning of specific GHG emission reduction targets. Other actions involve the conservation of green areas and energy efficiency. Adaptation strategies mainly involve the civil defense and urban planning sectors [17].

Coastal cities represent an important gap in Brazil's local climate policies, as they are considered even more vulnerable to climate change for a number of reasons. The first one is geographical specificity, followed by the interface between continent, atmosphere, and ocean. The third reason is the high concentration of people and infrastructures, which turns climate events into disasters since people and structures can be severely affected. Brazil has a coastline of almost 7500 km where some of the most important cities are located and where most of the population is concentrated. However, only five coastal cities (Fortaleza, Recife, Rio de Janeiro, Salvador, and Santos) have an adaptation strategy [17].

In Brazil's coastal cities, climate risks are already worsening the socio-environmental vulnerability of marginalized populations and creating real climate emergencies [6, 18–20]. In these areas, there is a greater risk of landslides, floods, coastal erosion, and heat waves [21, 22]. There are also high levels of socio-spatial segregation and many informal settlements in risk areas [23], besides institutional discontinuity and a lack of political representation [24] that contribute to the heightened risk in the country's coastal cities. An example of this was the recent tragedy of São Sebastião, in the coastal area of the state of São Paulo, when, on February 19, 2023, a storm caused landslides, carrying away houses and destroying some of the municipality's key infrastructures, and thousands were killed or went missing [25].

In terms of institutional mechanisms for policy implementation, 10 of the 12 cities have established a Climate Forum or Committee, with the participation of municipal secretariats and agencies, universities and research institutes, the private sector, and civil society organizations. **Table 1** offers a concise overview of the 12 municipal climate policies we identified in this research which combined deductive and inductive approaches [26] to analyze documents, observations, and semi-structured interviews [27]. To analyze the content, we employed a qualitative approach [26] and Structure Analysis [28]. Finally, we conducted a Multiscale Vulnerability Analysis [29]. The compiled results are presented in **Table 1** below ordered alphabetically by city.

In 2017, the "Frente Nacional de Prefeitos" (FNP – National Front of Mayors), the "Governos Locais pela Sustentabilidade" (ICLEI – Local Governments for Sustainability), and the European Union signed the Global Pact of Mayors for Climate and Energy, an agreement that promotes greater collaboration between the world's cities [30]. More than 120 Brazilian cities are part of this initiative, which sought to build connections between municipalities to increase the supply of financing and enable local actions for climate and renewable energy. It is considered the largest global alliance of cities and local governments to fight climate change currently.

City/ (state)	Total population (2023)[1]	Climate policy	Year	Mitigation strategies	Adaptation strategies	Institutional mechanisms of action
Belo Horizonte (MG)	2,315,560	Law 10,175	2011	30% reduction in GHG emissions by 2015	Adaptation plan in progress	Municipal Committee on Climate Change and Eco Economy (2006)
Curitiba (PR)	1,773,733	Decree 1186	2009	Mitigation plan in progress	Adaptation plan in progress	Curitiba Forum on Climate Change (2009)
Feira de Santana (BA)	616,279	Law 3169	2011	Goal to reduce GHG emissions, but no set target	To be defined	Municipal Forum on Global Climate Change and Biodiversity (2011)
Fortaleza (CE)	2,428,678	Law 10,586	2017	15.5% reduction in GHG emissions by 2020 and 20% by 2030	Adaptation plan under development	Fortaleza Forum on Climate Change (2015)
Manaus (AM)	2,063,547	Law 254	2010	Mandatory use of equipment aimed at the rational use of energy and water in buildings; and tax incentives for sustainable practices	Mandatory use of equipment aimed at the rational use of energy and water in buildings; and tax incentives for sustainable practices	Municipal government
Palmas (TO)	302,692	Law 1182	2003	Plan for conservation of green areas and energy efficiency	Not Defined	Municipal Secretariat of Environment
Porto Alegre (RS)	1,332,570	CL 872[2]	2020	GHG emission reduction targets to be set after inventory execution	Resilience Plan (2016)	Municipal Committee on Climate Change and Energy Efficiency (2016)
Recife (PE)	1,488,920	Law 18,011	2014	GHG emissions reduction plan with targets by sector of activity (2016)	Adaptation Plan (2019)	Recife Committee on Sustainability and Climate Change (Comclima) (2013), Executive Group on Sustainability and Climate Change (Geclima) (2013)
Rio de Janeiro (RJ)	6,211,423	Law 5248	2011	GHG emission reduction targets: 8% in 2012; 16% in 2016; 20% in 2020	Climate Change Adaptation Strategy (2016)	Rio de Janeiro Forum on Climate Change and Sustainable Development (2009)
Santos (SP)	418,608	Adaptation Plan	2022	Not Defined	Adaptation Plan (2016)	Municipal Committee of Climate Change Adaptation (2015)
São Paulo (SP)	11,451,245	Decree 60,290	2021	Climate Action Plan of the Municipality of São Paulo 2020–2050	Climate Action Plan of the Municipality of São Paulo 2020–2050	Executive Secretariat on Climate Change (2021)
Sorocaba (SP)	723,574	Law 11,477	2016	GHG emission reduction targets to be set after inventory execution	Adaptation plan in progress	Local Climate Change Committee and Working Group on Climate Change (2019)

Note: (1) Total population (2023): This column lists the total per capita population of a Brazilian municipality, expressed in millions; and (2) CL 872 is the Complementary Law no. 872 of the capital city of Porto Alegre, in the State of Rio Grande do Sul.

Table 1.
Climate change policies in Brazilian municipalities. Source: Updated from Barbi and Rei [17].

3. Climate policies at the Brazilian national level

Climate actions related to institutional political structures in Brazil at the national level can be divided into four phases.

- *First Phase* (1992–2002): The creation of the National Climate Change Program in 1994, the first political-institutional structure to respond to climate change, as well as the Commission for Sustainable Development and the Center for Climate Studies (INPE/CPTEC – Commission for Sustainable Development and the Center for Climate Studies). The Commission on Global Climate Change was created in 1999. During this phase, civil society played an active role. The Brazilian Forum on Climate Change, created in 2000, is a case in point. It brings together representatives from the government, civil society, the private sector and research institutes. The importance of civil society has been emphasized in the process with the work of the NGO "Observatório do Clima" (Climate Observatory) since 2002.

- *Second Phase* (2003–2008)*:* the development of a Climate Agenda takes shape. The first inventory of greenhouse gas emissions was conducted in 2004. The National Climate Change Plan started to be formulated in 2007 and concluded in 2008 along with the "Rede Global de Mudanças Climáticas" (Global Climate Change Network). It is worth noting the valuable contributions of the "Centro de Ciência do Sistema Terrestre" (CCST – Earth System Science Center) and the "Instituto Nacional de Ciência e Tecnologia de Mudanças Climáticas" (INCT - National Institute of Science and Technology of Climate Change) in 2008.

- *Third Phase* (2009–2012)*:* establishment of a Brazilian Climate Policy. The "Política Nacional de Mudanças Climáticas" (PNMC – National Climate Change Policy) took place in 2009 along with the National Climate Change Fund. The "Painel Brasileiro de Mudanças Climáticas" (PBMC – Brazilian Panel on Climate Change) was founded. In 2010, the second inventory of greenhouse gas emissions was published, and in 2012 the first PBMC report was released. This is a very significant period for the Brazilian process that laid the foundations for many of the policies on climate change that followed.

- *Fourth Phase* (2013–2023)*:* the implementation of the climate change policy begins with the following plans covering all Brazilian biomes and a wide range of sectors: A deforestation Control Plan for the Legal Amazon; Deforestation Control Plan for the Cerrado; 10-year Energy Plan; Low Carbon Agriculture Plan; Industry Transformation Plan; Low Carbon Emission Plan; and finally, the Transport and Urban Mobility Plan. In the middle of the decade, political resistance to several of the initiatives at the federal level formed and started to call into question the necessary emission cuts and reforms. This is where, similar to other federal systems, subnational initiatives began to show their importance for building continuity in Brazil's climate change policy.

While Brazil has played a key role in the construction of the global environmental agenda, in particular through the United Nations "Rio Conventions" [31], their implementation in the country has had mixed success. Some of the backlashes have to do with powerful national industry groups like the agricultural lobby or a number of recent fossil fuel discoveries that have impacted important political narratives about climate change [32]. This means there have been clear steps backward, for instance when the state-owned petroleum company Petrobras increased its production of oil and other fossil fuels; GHG emissions have logically also increased, and connected deforestation has increased since 2012. Given the context of what we describe here as "one step forward, two steps back", the most effective way to respond to the climate emergency is to focus on initiatives at the state and local levels for example, in the energy transition.

What is needed is a reduction in emissions, since the volume has only increased in Brazil since 1992. The results of the drastic political shift toward the right in 2015 and the five successive conference of the parties (COPs) appear to support this view, as there has been no commitment from Brazil to immediately reduce emissions. It seems that future targets have been set that may prove challenging to meet, at least in the short term.

4. Conclusions

In this chapter, we have documented the evolution of Brazilian climate change policies and the importance that the inclusion of different social actors has had over the last two decades.

These changes in Brazil's climate governance have had significant impacts on the resilience and continuation of the adaptation and mitigation agenda. Especially the increasing activities of subnational actors stand out in an increasingly polarized political environment at the federal level. Since 2019, there has been open skepticism about climate change and a lack of concern about its risks in the policies of the then-president Jair Bolsonaro. Furthermore, the federal government has encouraged actions often not aligned with environmental protection, protecting the interests of the "rural caucus" leading to an increase in deforestation and the burning of natural resources.

Upon review of the existing policies, there seems to be a discrepancy between the existence and the fundamental purpose of the created institutional structures. This calls into question the actual achievement of objectives set in national agreements, agencies, and plans as opposed to (often ratified) international commitments. This discrepancy is evidenced by the increase in the rising consumption of fossil fuels, continued deforestation rates, and the annihilation of biodiversity even before the presidential term of Bolsonaro [33, 34].

This discrepancy or lack of alignment between federal and local climate policies may indicate an opportunity for improvements in the governance structure. Some Brazilian analysts have called this phenomenon "disgovernance" - a sensitive and crucial point in the country's efforts to address climate change. This becomes clear in two Brazilian publications we would like to highlight here.

The first was published in Revista FAPESP under the title 'The world boils" in 2023 and explains some of the back-and-forth of climate "(dis)governance" *[(des) governança* in Portuguese) [35]. The study pointed to a planet Earth on a heating course without brakes, in which global warming caused the hottest month in the last 150 years, accentuating the climate crisis. At the time of publication, the global temperature reached 20.96 degrees Celsius, making 31 July 2023 the hottest day in the planet's recent history and certainly, this record may already be broken by the time of publication. The second article, entitled "Climate, Justice and Governance", was published in the Jornal da Unicamp, presenting data and information on the role of universities as mediators in this process [36]. While important steps have been taken in the reconstruction of environmental institutions, the country's leadership seems unwilling to move away from a development model that does not let go of fossil fuels. In their reflection on the outcomes of an academic seminar, the authors lament this mentality and ask for a more proactive role of universities in questioning a development model that further aggravates the climate crisis instead of contributing to solving it.

In the Brazilian context, extreme weather events are already being witnessed: record droughts in the Amazon [37], fires in the Pantanal swamplands [38] severe floods in Rio Grande do Sul. In the latter, an unprecedented calamity was caused in which more than 270 of the state's 497 municipalities and the state capital were under water during the month of April [39]. At the same time heat waves swept across the country. This urgently calls for structural resilience solutions that go beyond a merely reactive approach time and again.

It might be helpful to explore additional strategies for addressing land grabbing, burning, and deforestation of green areas across the country, which are largely driven (and even politically promoted) by agribusiness and the expansion of urban areas. Additionally, there is a growing need for urban development that enhances the resilience, safety, and inclusivity of cities in the face of disaster risks.

It is clear that Brazil's fragile governance system—the result of a political arena that has been degrading and weakening the country's environmental agenda—needs to be addressed as a whole. There is a clear and pressing need for this process to be reversed, integrating diverse interested parties into the complex political game of public climate and environmental policies at all levels. This should be done as outlined by Seixas et al. [36], through a participatory dialog.

It is evident that at the national level, the country still has a significant amount of work to do to advance its environmental policies. Since 2012, many of these policies have been discontinued. Since 2018, they have been met with structural resistance from the federal level. To mitigate deforestation and land use and occupation that increases greenhouse gas emissions, we must promote integrated public policies and strengthen institutions that have been created for this purpose since Brazil's democratization in 1988.

The following efforts are recommended:

1. Promotion of the National Council of Environment (Conama) and its participatory chambers;

2. Reinforcement of the "Brazilian Institute for the Environment and Natural Resources" (IBAMA) and the "Chico Mendes Institute for Biodiversity Conservation" (ICMBio) with financial, technical, and human resources to enable them to effectively inspect and prevent environmental crimes in Brazil;

3. Review of the purchase process of land's important biomes (Amazon, Cerrado, Atlantic Forest, Pantanal, Pampa, etc.).

For this purpose, blocking bills in the Federal Chamber of Deputies that allow or facilitate this kind of business, such as the proposed law 2963/2019, is an urgent measure. Finally, as a lot of the defunding of environmental monitoring has been justified with necessary reductions in public spending, the national campaign for the

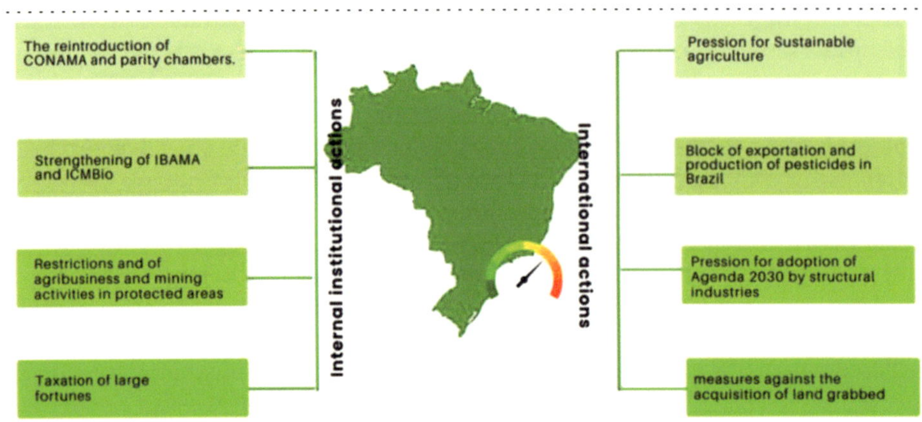

Figure 2.
Some recommendations for climate governance in Brazil. Source: Made by the authors.

taxation of large fortunes should become a reality. The amount collected from the tax could finance climate adaptation and mitigation works in cities impacted by current extreme events. **Figure 2** exposes a diagram of our suggestions.

On the other hand, international reinforcement is as welcome as it is necessary. In this respect, we consider essential:

1. International pressure for Brazil's commodity agribusiness to review its production and marketing patterns in favor of more sustainable agriculture. Only in this way, the country can stop relying on an unsustainable development mode as one of the biggest exporters of cheap natural resources to the world economy [water, solar energy, biodiversity, food, etc.] while not providing these to its own population;

2. Drastically reducing the export of pesticides to Brazil;

3. Demanding that Brazilian companies effectively adopt the 2030 Agenda, especially national and foreign corporations in the chemical, mining, energy, infrastructure, and transport sectors, and;

4. Measures against the acquisition of land grabbed by foreign business groups in the country.

The risks posed by global climate change are affecting Brazil in a number of very visible ways already. They affect everyone, in the Global North and the Global South, and it is becoming increasingly difficult to "buy our way out" of this crisis. In line with Beck's [2] final ideas about the positive societal forces the enormous risks of climate change can generate, our hope is that organized Brazilian civil society will ultimately manage to hold its political representatives more accountable, maintain and even expand its much-needed representation in the coordination of solutions to this crisis – and finally make two steps forward again.

Author details

Leila da Costa Ferreira[1*], Fabiana Barbi[2], Niklas Weins[3] and Luciana Lima Domingues de Souza[4]

1 Sociology Department, University of Campinas, Campinas, Brazil

2 University of São Paulo, São Paulo, Brazil

3 Xi'an Jiaotong Liverpool University, Suzhou, China

4 University of Campinas, Campinas, Brazil

*Address all correspondence to: leilacf@unicamp.br

References

[1] IPCC, 2021: Summary for policymakers. In: Masson-Delmotte V, Zhai P, Pirani A, Connors S L, Péan C, Berger S, et al editors. Climate Change 2021: The Physical Science Basis. Contribution of Working Group I to the Sixth Assessment Report of the Intergovernmental Panel on Climate Change. Cambridge University Press, Cambridge: United Kingdom and New York, NY, USA; pp. 3-32. DOI:10.1017/9781009157896.001

[2] Beck U. A Metamorfose do Mundo. Novos conceitos para uma nova realidade. Zahar Ed: Rio de Janeiro; 2018

[3] Dunlap R, Brulle R. Climate Change and Society. Sociological Perspectives. New York: Oxford University Press; 2015

[4] Ferreira LC. The multi-level governance challenge of climate change in Brazil. In: Bell M, Carolan M, Legun K, Keller J, editors. The Cambridge Handbook of Environmental Sociology. Cambridge: Cambridge University Press; 2020

[5] Ferreira LC, Barbi F, Barbieri M, organizers. Dimensões Humanas das Mudanças Climáticas no Sul Global. Curitiba: Fapesp. CRV Editora; 2020

[6] Ferreira LC, Seleguim FB, organizers. A emergência climática: governança multinível e multiatores no contexto brasileiro. Curitiba: CRV/FAPESP; 2023

[7] Betsill MM, Bulkeley H. Looking back and thinking ahead: A decade of cities and climate change research. Local Governments. 2023;**12**(5):447-456

[8] Carbon Brief. Analysis: Which Countries are Historically Responsible for Climate Change? [Internet]. 2021. Available from: https://www.carbonbrief.org/analysis-which-countries-are-historically-responsible-for-climate-change/

[9] SEEG. Emissões Totais [Internet]. 2022. Available from: http://plataforma.seeg.eco.br/total_emission# [Accessed: 22 June 2022]

[10] Maluf RS et al. Global value chains, food and just transition: A multi-scale approach to Brazilian soy value chains. The Journal of Peasant Studies. 2022;**50**(7):2642-2665. DOI: 10.1080/03066150.2022.2105700

[11] University of Notre Dame. Notre Dame Global Adaptation Initiative (ND-Gain) [Internet]. 2023. Available from: https://gain.nd.edu/our-work/country-index/ [Accessed: 19 August 2024]

[12] World Bank. Climate Risk Profile: Brazil. The World Bank Group [Internet]. 2021. Available from: https://climateknowledgeportal.worldbank.org/sites/default/files/2021-07/15915-WB_Brazil%20Country%20Profile-WEB.pdf [Accessed: 20 August 2024]

[13] Schneider T, Betti P, Ferreira CJ, Julião DP, Silva SJG. Guia de adaptação e resiliência climática para municípios e regiões. São Paulo: Secretaria de Infraestrutura e Meio Ambiente (SIMA); 2021

[14] World Economic Forum. The Global Risks Report 2022. 17th ed. Cologny/Geneva, Switzerland: World Economic Forum; 2022

[15] Instituto Brasileiro de Geografia e Estatística (IBGE). Panorama do Censo

2022. Available from: https://censo2022.ibge.gov.br/panorama/ [Accessed: 24 August 2023]

[16] Barbi F, Ferreira LC. Climate change in Brazilian cities: Policy strategies and responses to global warming. International Journal of Environmental Science and Development. 2013;**4**(1):49-51

[17] Barbi F, Rei F. Federalism and climate governance in Brazil. In: Jodoin S, Setzer J, organizers, editors. Climate Change and Federal Governance. Cambridge: Cambridge University Press; 2023

[18] PBMC. In: Marengo JA, Scarano FR, editors. Impacto, vulnerabilidade e adaptação das cidades costeiras brasileiras às mudanças climáticas: Relatório Especial do Painel Brasileiro de Mudanças Climáticas. Rio de Janeiro, Brasil: PBMC, COPPE – UFRJ; 2016

[19] Barbi F, Ferreira LC. Risks and political responses to climate change in Brazilian coastal cities. Journal of Risk Research. 2013;**17**(4):485-503. DOI: 10.1080/13669877.2013.788548

[20] Baptista SP. Cidades e mudanças climáticas: desafios para os planos diretores municipais brasileiros. Dossiê: a metrópole e a questão ambiental. Cadermps da Metrópole. 2020;**22**(48):365-394. DOI: 10.1590/2236-9996.2020-4802

[21] Harari et al. Sensing sociability: Individual differences in young adults' conversation, calling, texting, and app use behaviors in daily life. Personality and Social Psychology. 2020, 2019;**119**(1):204-228. DOI: 10.1037/pspp0000245

[22] Souza et al. Revista Vértices. 2020;**22**(3):590-609. DOI: 10.19180/1809-2667.v22n32020p590-609

[23] Oscar Junior et al. Emergência climática: desafios e oportunidades no campo do ensino de geografia. Revista da Anpege. 2019;**18**(36):713-739

[24] Di Giulio et al. Eventos extremos, mudanças climáticas e adaptação no Estado de São Paulo. Ambiente and Sociedade São Paulo. 2019;**22**:e02771

[25] G1 Vale do Paraíba e região. Tragédia de São Sebastião: veja como estão os sobreviventes um ano após o temporal devastador [Internet]. 2024. Available from: https://g1.globo.com/sp/vale-do-paraiba-regiao/noticia/2024/02/19/tragedia-de-sao-sebastiao-veja-como-estao-os-sobreviventes-um-ano-apos-temporal-devastador.ghtml [Accessed: 19 August 2024]

[26] Mayring P. Qualitative content analysis. Forum Qualitative Sozialforschung/Forum: Qualitative Social Research [Internet]. 2000; 1, n.2, 2000. DOI: 10.17169/fqs-1.2.1089. Available from: https://www.qualitative-research.net/index.php/fqs/article/view/1089 [Accessed: 20 August 2024]

[27] Flick U, Von Kardorff E, Steinke I. Qualitative Forschung. Ein Handbuch. 14 ed. Reinbek: Rowohlt Taschenbuch, 2000

[28] Entman. Framing: Toward clarification of a fractured paradigm. Journal of Communication. 1992;**43**(4):51-58. DOI: 10.1111/j.1460-2466.1993.tb01304.x

[29] Iwama et al. Risco, vulnerabilidade e adaptação às mudanças climáticas: uma abordagem interdisciplinar. Ambiente e Sociedade. 2016;**19**(2):95-118. DOI: 10.1590/1809-4422ASOC137409V1922016

[30] Global Covenant of Mayors. Global Covenant of Mayors stresses the importance of Climate Adaptation Plans

during the Climate Federalism Workshop in Brazil. 2024. Available from: https://pactodealcaldes-la.org/language/en/global-covenant-of-mayors-stresses-the-importance-of-climate-adaptation-plans-during-the-climate-federalism-workshop-in-brazil/

[31] Dauvergne P, Farias D. The rise of Brazil as a global development power. Third World Quarterly. 2012;**33**(5):903-917. DOI: 10.1080/01436597.2012.6747

[32] Viglio JE, Di Giulio GM, Barbi F, Ferreira LC. Narrativas científicas sobre petróleo e mudanças do clima e suas reverberações na política climática brasileira. Sociologias. 2019;**21**(51):124-158. DOI: 10.1590/15174522-0215105

[33] Levis C et al. Help restore Brazil's governance of globally important ecosystem services. Nature Ecology and Evolution. 2020;**4**(2):172-173. Springer Science and Business Media LLC. DOI: 10.1038/s41559-019-1093-x

[34] Souza L et al. Crise ambiental e a desnaturação da Política Nacional do Meio Ambiente no Brasil, Reverse Geocoding. 2021;**20**(1):12-35. DOI: 10.33947/1981-741X-v20n1-4785. Available from: https://revistas.ung.br/index.php/geociencias/article/view/4785 [Accessed: 21 August 2024]

[35] Marques F, O mundo mais quente. Relatório do IPCC amplia grau de confiança sobre diagnóstico das mudanças climáticas. Revista FAPESP, N. 212. 2023. pp. 39-41. Available from: https://revistapesquisa.fapesp.br/o-mundo-mais-quente/ [Accessed: 20 August 2024]

[36] Seixas SRC, Ferreira LC, Thezolin R, Jornal da UNICAMP. 2023; Era da Fervura Global: os desafios de governança e de justiça climática [Internet]. Available from: https://unicamp.br/unicamp/ju/artigos/ambiente-e-sociedade/era-da-fervura-global-os-desafios-de-governanca-e-de-justica/ [Accessed: 20 August 2024]

[37] Jornal da UNESP. Seca que afetou a Amazônia em 2023 causou a maior queda nos níveis dos rios já registrada, e está relacionada a mudanças climáticas, mostra estudo [Internet]. 2024. Available from: https://jornal.unesp.br/2024/04/24/seca-que-afetou-a-amazonia-em-2023-causou-a-maior-queda-nos-niveis-dos-rios-ja-registrada-e-esta-relacionada-a-mudancas-climaticas-mostra-estudo/ [Accessed: 20 August 2024]

[38] Camara J/g1 MS. Pantanal: incêndio de 2024 supera o registrado no mesmo período de 2020, ano recorde de queimadas [Internet]. 2024. Available from: https://g1.globo.com/ms/mato-grosso-do-sul/noticia/2024/06/20/pantanal-incendio-de-2024-supera-o-registrado-no-mesmo-periodo-de-2020-ano-recorde-de-queimadas.ghtml [Accessed: 20 August 2024]

[39] g1 RS. Um mês de enchentes no RS: veja cronologia do desastre que atingiu 471 cidades, matou mais de 170 pessoas e expulsou 600 mil de casa [Internet]. 2024. Available from: https://g1.globo.com/rs/rio-grande-do-sul/noticia/2024/05/29/um-mes-de-enchentes-no-rs-veja-cronologia-do-desastre.ghtml [Accessed: 20 August 2024]

Chapter 8

A Review of Residual Flood Risks in South African-Vulnerable Coastal Communities: Opportunities to Influence Policy

Indrani Hazel Govender, Maliga Reddy and Rajendran Perumal Pillay

Abstract

In recent years, many parts of South Africa have been devastated by floods, impacting severely on the most vulnerable communities. Despite measures to reduce flood risk, and implementation of adaptation measures, there is always a measure of harm to livelihoods, health and wellbeing, and the economy, which persist far beyond flood events. This is of particular concern as flood events have increased in frequency and severity. A review of literature, addressing flood risk management, mitigation and adaptation in vulnerable coastal communities of South Africa, was conducted, to determine how to manage residual risks through implementation of resilience and recovery programs, employing suitable tools, inclusive of appropriate multidisciplinary stakeholders. This study interrogated different approaches used to assess social vulnerabilities, risk perception and the role of risk communication and information dissemination. Economic implications pertaining to tourism, livelihoods and loss of natural and built infrastructure were analyzed. Risk tools including early warning systems and risk assessment models were analyzed, and various future scenarios were explored. Possible opportunities presented by negative impacts of residual risks to achieving the SDGs were explored. The findings indicate that strengthening resilience depends heavily on collaboration across sectors to cater for local needs. Cooperation between government, private sector and communities is critical to achieving sustainable solutions to residual risk management.

Keywords: flood risk management, residual risks, vulnerable communities, climate adaptation, policy development, coastal flood risks

1. Introduction

Climate change combined with human activities and land transformation poses considerable risk to human health and wellbeing, ecosystems and the economy.

Risk may be broadly defined as "the product of a hazard and its consequences" [1]. In the absence of harm to entities that are valued, such as people, property or natural areas and biodiversity, there is no risk. In the disaster risk landscape in South Africa, the largest losses are due to climate risk [2]. Climate risks impact negatively on economic activities, access to educational facilities, water quality and supply, and infrastructure [3, 4]. In the last two to three decades, we have seen a surge in extreme weather events (EWE), with floods topping the list as the most recurrent natural disaster [5]. This has resulted in the transformation of large areas, as recovery and restoration to the original condition are not entirely possible. We can only prepare adequately for what we anticipate with certainty. However, EWEs associated with climate change are linked to overwhelming uncertainty.

Flood hazard considers "the likelihood and severity of flood events" [6]. Globally, Africa and Asia are most impacted by climate change, contributing to food security and forced migration [4]. Africa has been identified as the continent worst impacted by climate change, where "one in every three people lives below the poverty line" [7]. EWEs may introduce wetter than normal conditions which have impact on food safety, health and well-being and livelihoods. South Africa has been subjected to damaging floods approximately every 2 years, with more intensive flood events occurring every 10–15 years but the frequency of the latter event is expected to increase [8–10]. South Africa has experienced more frequent flood events over the past few decades, with 77 major floods occurring between 1980 and 2010, resulting in the loss of over 1000 lives, devastation to infrastructure and loss of livelihoods [6]. Between 1950 and 2021 South Africa experienced 40 disastrous floods, with the east coast subjected to repeat flood events in the last three decades. The floods experienced in KwaZulu-Natal, South Africa, in 2019 left in its wake 80 fatalities, 1400 displaced people and damage estimated at ZAR140 million, while the April 2022 floods resulted in damages estimated at ZAR17 million [11].

1.1 Coastal settlements vulnerable to risk

Coastal towns and cities are especially vulnerable to the impacts of climate change and associated EWEs, particularly floods, and sea level rise (SLR) [12, 13]. SLR is not uniform and becomes more pronounced through discrete events that occur together with severe storms [14]. Coastal areas have become a concern due to the increasing flood events in coastal areas [8]. Vulnerability may be defined as "the lack of resistance to damaging/destructive forces" [1]. This is of particular concern in coastal areas due to the dynamic nature of natural forces from both the sea and land. As rivers are inundated with extreme volumes of water, and move sediment and debris downstream, the ultimate destination is the coastal area. With many settlements clustered closer to the coast, the high-density populations are recipients of the dire consequences of these high-risk events. Coastal floods also result from "high tides, storm surges and strong winds," river floods and flash floods [1, 6]. The combination of inland flooding and coastal flooding can have considerable impact on coastal settlements. The magnitude, duration and frequency of flood events have become increasingly unpredictable, resulting in inadequate preparation for such events [1, 15]. In most instances, the intensity of floods is increasing, with resultant loss of lives and property, and devastating social and economic costs.

In developing countries, poverty and a multitude of socioeconomic and political factors contribute to placing vulnerable communities in a further precarious position.

Informal settlements are permanent homes to approximately one billion people in the Global South [16, 17]. Basic services such as water and sanitation, solid waste removal, and adequate drainage are the norm in these settlements [18]. Due to the lack of services, informal settlements are often located along water sources and in drainage areas, which increases their vulnerability to flood events. These communities are often recurrent victims of floods, as they rebuild their unstable dwellings after each flood event and continue to reside in the same area. They are characterized by overcrowding, unemployment and low income [18–20]. Apart from informal settlements in South Africa, environmental injustice stemming from the historical political dispensation resulted in residential settlements being segregated, with certain groups residing in areas prone to floods. These were also the groups that faced socioeconomic challenges and limited access to resources and services. It is also noted that more settlements are arising in hazard-prone areas as time progresses [21], as people flock to urban areas.

Cape Town and Durban are the largest coastal cities in South Africa, with each city having a population exceeding 3 million people [20]. The Climate Change Act No. 22 of South Africa [22] in its principles acknowledges "the need for decision-making to consider the special needs and circumstances of localities and people that are particularly vulnerable to the adverse effects of climate change." This is especially important in coastal regions, with a high concentration of the population, with many leaving inland areas to work at the coast. Many of these migrant workers settle in informal dwellings, which place them in harm's way when flood events occur.

1.2 Residual flood risk

Floods are a regular occurrence but the uncertainty in terms of predicting when they will occur and to what extent is challenging. Using a risk assessment approach is often effective in identifying hazards (causes) and effects (consequences) [6]. Risk assessment and probability of consequences under different scenarios help in planning for mitigation and adaptation [4]. However, there is always some degree of negative consequences that are unexpected, leaving communities and the environment at the mercy of flood events. This is residual harm or residual risk that is anticipated harm. Residual risk may be defined as "the risk that remains after the implementation of protection measures" [23] or "risk that remains in unmanaged form, even when effective disaster risk reduction measures are in place, and for which emergency response and recovery capacities must be maintained" [24]. Dealing with recovery following flood events has required considerable investment but many communities have failed to fully recover and return to what used to be their "normal" lives. This is a function of the vulnerability to flood event, which considers livelihoods, people and infrastructure [6].

This chapter aims to explore the management of residual flood risks to determine their impact on vulnerable communities in coastal areas of South Africa, through a review of flood risk management, mitigation and adaptation literature. This review examines how resilience and recovery interventions can address residual risks through the application of suitable approaches and tools. The challenges that residual risk poses to achieving the sustainability agenda, including specific barriers to successful implementation of appropriate mechanisms to manage residual risk, are examined. Ultimately, the implications of these significant issues for policy development are analyzed and recommendations are proposed.

2. Methodology

A literature search of peer-reviewed publications was conducted for the period 2010 to 2024, limited to South Africa. Search terms used were: "residual risk in coastal regions"; "coastal flood risk assessment" AND "coastal flood risk management in South Africa"; "coastal flood vulnerability in South Africa"; "climate change impact on coastal flooding in South Arica"; "socio-economic impacts of flood in South Africa." The databases used to conduct the literature search and the results generated appear below in **Figure 1**.

- A Google Scholar search generated 23 articles after refining by year and country. After removal of duplicates, 19 were found to be eligible.
- ScienceDirect was used to conduct a literature search. After refining by year, document type, language, keywords and subject area, 22 were included, and 14 refined to be eligible after screening.
- A literature search using Scopus generated 16 articles, after refining by year and country, 11 were found to be eligible.

A total of 44 publications generated from the search above. These were screened further for relevance, and the final review included 34 publications. The content of the selected literature was examined within the broad themes: (1) social vulnerabilities of communities, health and wellbeing and displacement, (2) economic vulnerabilities

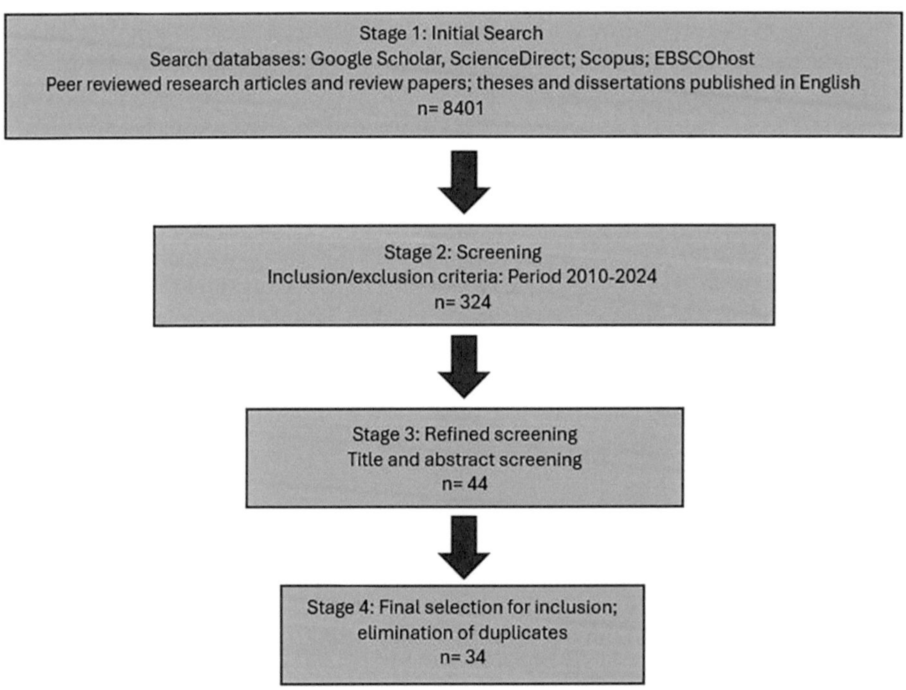

Figure 1.
The stages in the literature search and refinement to determine the final selection for inclusion in the study, commenced with 8401 publications. This was screened for the period 2010 to 2024 generating 324 publications and refined further to include 34 publications in the study.

and broader implications, (3) risk tools used to assess vulnerabilities and (4) risk scenarios and consequences. This was reviewed within the broad context of the ultimate contribution toward policy development.

Apart from this, other publications deemed appropriate were used to broaden the review where possible in the discussion section, as certain highly relevant publications were not part of the literature selected for the review but had a strong bearing on the aim of this chapter. This allowed the authors to make further inferences.

3. Results and discussion

Between 2014 and 2021, there were between 0 and 2 publications per year. There was a sharp increase in the number of relevant publications retrieved from 2022 to 2024 from the literature search conducted. This may be attributed to the severe and devastating floods experienced in 2022–2023 [9] (**Figure 2**).

3.1 Social vulnerabilities

Climate change has serious impacts for humankind with coastal communities among the most vulnerable [10, 12, 25, 26]. Social vulnerability is understood as a dynamic component changing over time, raising concerns of the inability of a population or community to cope with hazards such as floods [27]. These susceptibilities are often grouped as socioeconomic vulnerabilities. Social vulnerability can be typified by factors such as age, income, physical disabilities or other medical conditions, medical and health care access, well-being, gender, race, and ethnicity, access to resources, building infrastructure, population density and the nature and intensity of the hazard itself [10, 27, 28]. The literature provides evidence that fatalities may have reduced locally and globally, but social and economic repercussions remain a challenge in flood risk management [26]. Busayo et al. [25] report that studies in South Africa seem to focus primarily on physical risks and damage, with far less emphasis on social vulnerability of those most affected by flooding. In addition, South Africa has the

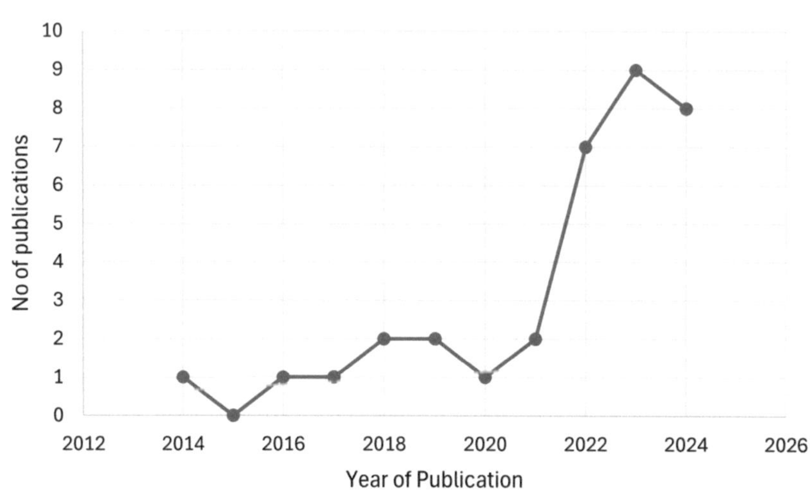

Figure 2.
The number of publications per year over the period covered for the review, 2010 to 2024, was low ranging from 0 to 2 from 2014 to 2021. From 2022, there was a steep increase, with a maximum of nine publications in 2023.

legacy effects of apartheid resulting in "high risk areas and poverty traps where high rates of losses and damages go hand in hand with limited ability to cope and adapt due to marginalization, high poverty, and culturally imposed gender roles" which exacerbate the impacts of climate change [20].

3.1.1 Different approaches to understanding and assessing social vulnerability

Of the 34 publications reviewed, 11 papers refer to social vulnerabilities, while the other publications acknowledge a social component indirectly. One of the criteria on the Joanna Briggs Institute (JBI) of qualitative analysis is: Are participants, and their voices, adequately represented [29]? This is critical to understanding socially vulnerable communities due to flooding. A vital component to understand social vulnerabilities to both mitigation and response, is the lived experiences for both pre- and post-flood events. Four papers make categorical references to the voices of the vulnerable. These are: Busayo and Kalumba [28] reported on a study in East London, a coastal city in South Africa, Williams et al. [17] reported on Quarry Road West informal settlement, in Durban, along the east coast of South Africa, Quinn et al. [30] reported on residents' responses on place meaning (an aspect of social vulnerability component) in four coastal towns in France, South Africa and UK and two French towns, and Ebhuoma et al. [10] reported on household preparedness for interventions and flooding along the coastal town of Amanzimtoti, South Africa. Ebhuoma et al. [10] explored the lived experiences of coastal vulnerable communities using a case study. Other publications, while not categorically focusing on the voices of the vulnerable, suggested approaches, which could contribute to understanding and mitigation of social vulnerabilities within the context of coastal flooding events and residual risk. These include: the GULLS model (Global Understanding and Learning for Local Solutions) [12], the Safe Development Paradox [31], an ecosystem-based adaptation (EbA) approach [25], green infrastructure (GI) as a mitigation strategy [19], understanding rural and urban contexts [32], an integrated approach to flood resilience in urban contexts [33], adaptive strategies of communities in coastal inundation drawn from literature review [34] and contributions of social engagement relating to expansion from flood control [35].

Aswani et al. [12], in citing many authors, aver that vulnerability assessment and adaptation strategies are dual factors that are essential for holistic risk management. They refer to the GULLS model and the application of this conceptual model to vulnerability analysis. The GULLS project was implemented along coastal hotspots in countries across the globe (Australia, Brazil, India, South Africa, Madagascar and Solomon Islands) where warming of coastal waters has been faster. This model gathers local social vulnerability data to obtain a better understanding of drivers of community vulnerabilities. While this is applied at a local scale, it allows for up-scaling to a regional level. Understanding local cultures and knowledge has been raised in other papers as well, namely by Busayo et al. [25], Tauhid and Zawani [19], Zhou et al. [32] and Williams et al. [17].

A point raised in data collection in the GULLS Project was the use of translators to access the views of coastal communities of varying cultures [12]. In South Africa, five different communities within the same coastal region formed the study site. However, only one translator was used with bias to the Afrikaans-speaking population, since the study sites were mostly Afrikaans-speaking communities. While this in itself highlights the importance of understanding the communities, it raises caution that inclusivity is crucial in understanding vulnerabilities where there is diversity. Hence, in

South Africa a single translator for one language will be deemed insufficient for social vulnerability data collection given the multicultural and multilingual contexts. It is the view of the reviewers that the lack of a comprehensive understanding of cultural contexts is likely to have poor responses in the future in terms of social vulnerability, residual impacts and damages. However, a contribution emanating from the GULLS project is the value of creating a cultural database of local demographics, which can be useful to practitioners to access for both mitigation and response to flood events.

Safe development in the form of infrastructure lock-ins is counter measures to flooding [31]. However, safe development has become paradoxical and has raised concerns as a mitigation strategy. These structures may not withstand EWEs, posing further risks. While these structures are conceived and built by local policymakers as a major defense against hazards such as flooding, Busayo et al. [25] argue that such structures are unsustainable without considering ecological-based development. An isolated engineering approach needs a critical rethink by policy makers, especially where social vulnerability is a major concern. Flooding is increasingly accounting for natural disasters. Germany (2021) and Spain (2024) displayed one of the more recent breaches in flood defenses. Global statistics show that the annual average death casualties ranged between 5000 and 15,000 persons [31]. The social vulnerability of death and the impact on family structures in surviving generations could be immeasurable. The literature reviewed also shows that the Safe Development Paradox has been inadequately studied in the global South, in particular in socially vulnerable communities.

Safe development also places greater risks by settling populations in areas that are prone to such extreme hazards providing false securities [31]. A review of safe development infrastructure and current studies on the risks involved is required. For example, bridges that may have been thought to be designed with resistance, to mitigate social vulnerabilities need to be revisited from the engineering perspective for added security. Collapsed bridges had significant impact on the KwaZulu Natal (KZN) South Coast, South Africa, causing commuting challenges in getting to work and in education facilities. Learners lost valuable instruction time due to flooding; there were also disruptions experienced in the National School Exit Examinations in the Eastern Cape South Africa in 2024. Learning from the Global North is a positive contribution to the Global South as they develop policies for mitigation and adaptation. However, a key point raised is that there are limitations in flood infrastructure design; hence, household adaptation and resilience require engagement and dialog between communities, governments and the wider public to overcome false securities. Social vulnerability is often hidden where science-based solutions are taken as resistant buffers creating a shift to a passive paradigm rather than active resilience in communities [26].

3.1.2 Relocation of vulnerable communities to mitigate against residual risks

A suggested flood mitigation strategy to counter the Safe Development Paradox is relocation [31]. However, while relocation may be a recommendation in the Global North, it may not necessarily apply to highly populated coastal communities with cultural attachments, especially in coastal towns and cities of the Global South. Relocation may require people to travel great distances to work and creates challenges for accessibility [19]. Further, relocation would mean spatial and infrastructure development, which implies transferring the issue to areas less likely to experience the hazards. In reality climate change, EWEs have shown the contrary; any area could be susceptible to the impacts of such natural events. However, responses for temporary relocation depending on the seriousness of the hazard may be an option to both physical damage and social vulnerability, that is, to keep family units intact, to make available medical assistance and food to identified holdings (relocations). Prior communication of such holdings for temporary safety should be given to communities in preparation for such unanticipated EWEs to make temporary or permanent relocations a fairly good response. Identified relocations are also likely to help practitioners in times of EWEs to focus their responses. This would imply that a central database of vulnerable communities together with areas of temporary relocation needs to be available in advance to practitioners for acute awareness and response.

Perhaps a mitigation strategy to buffer flood hazards in conjunction with safe development is the ecosystem-based adaptation (EbA) approach advocated by Busayo et al. [25], which claims EbA to be cost-efficient and multi-dimensional and reduces environmental impacts. They indicate that housing shortages have resulted in mushrooming informal urban settlements exposed to flood risks in coastal areas such as in the Western Cape (South Africa). A hybrid approach of engineering and environmental management indicators (protection of ecosystems, conservation of coastal dunes, sustainable agriculture by preventing vegetation depletion along riparian areas and restoration), together with early warning systems and information dissemination, are mitigation measures, which increase preparedness and are very likely to decrease the impact on social vulnerability. A criticism though is that theoretically, it is an assumption that EbA will include the vulnerable communities first-hand if implemented appropriately. However, the EbA approach is likely to have a big gap if communities at risk are not involved. It is important to establish how people perceive EbA and associated ecological infrastructure, so that how people value such interventions is understood [36]. The Green Infrastructure (GI) approach to flood mitigation is similar to the EbA approach [19]. GI encompasses the establishment and development of interconnected networks of green spaces that sustain natural ecosystems and have multiple social, economic and environment benefits including slowing or controlling flood water run-off. It is acknowledged that GI has been implemented but issues related to the poor urban populations are rarely given the attention deserved [19]. Social vulnerabilities as evident in other cases include disability, water-borne diseases and death. Three key aspects of GI programs relevant for coastal flood mitigation and reduction in residual risks are awareness, participation and a shared vision, which are reported to have contributed to the success of GI programs in Madurai, India. However, the case studies lack evidence of the lived experiences of the poor.

3.1.3 The role of integrated knowledge systems

Bwambale et al. [26] propose the integrated knowledge approach (science and indigenous knowledge) that seeks to narrow the science-practice gap. They refer to hylomorphism (context-specific lived experiences integrated with the science that models the patterns of events). Kamara et al. [37] within the context of drought responses indicated that resilience capabilities and potential were evident in communities with loaded indigenous and contextual ecological knowledge. Indigenous knowledge is core to the lived experiences for traditional communities within contextual ecosystems including their comprehension of flood hazards [26]. However, the question is: How is indigenous knowledge or elements of the lived experience going to be cascaded to future generations with the science? Bwambale et al. [26] point out three issues: (1) a dearth of case studies of integrated knowledge, (2) arguments that even with integrated knowledge systems there is residual risk and (3) none-the-less there are co-benefits of communities being included in the policy and decision-making. The point here is significant if lived experiences include indigenous and other communities of urban, peri-urban and rural communities along flood risk coastal areas. For example, extreme flooding along the KZN South Coast is no longer a rare event but has become repetitive. Lived experiences are crucial to input on the impact of social vulnerability, mitigation and adaptative responses. However, policy and recommendations need to be drafted and communicated with language and terminologies understood by the communities at risk [26] inclusive of relevant traditional knowledge. Strengthening resilience includes the use of traditional and indigenous knowledge, culture and inherent capacities of communities at risk [28, 37].

Kamara et al. [37] reported on wellbeing due to climate induced disasters. Wellbeing is defined as a reasonable state without physical ill-health. During hazardous events, humanitarian aid is based on emergency provisions. However, building resilience should be a pre-condition. Poverty is one of the social vulnerabilities, which many nations are battling with, including South Africa, with its prevailing inequalities. Mudefi [11] reported that along the Palmiet River, in KZN perpetual disparities (e.g., unemployment and poverty) had negative impacts forcing Black South Africans to live in flood-prone informal settlements. Households with limited access to resources are disadvantaged and less resilient, increasing vulnerability to ill health and food shortage [37]. There seems to be inadequate studies and information on contextual resilience capabilities to inform policies. Lack of local government action, for example, Ethekwini Municipality in the Palmiet River settlement, failed to clear blocked drains, which increases flooding risks. Poor maintenance reduces resilience and increases social vulnerability. Accordingly in the 2022 floods experienced in KZN, South Africa, schools, hospitals and clinics, the core fabric of social wellbeing was eroded [11]. However, the review by Mudefi [11] lacked emphasis on the lived experiences of the communities at risk. The focus slanted toward responsible and accountable units of municipalities, government and other organizations. However, Muyambo et al. [38] have indicated that due to poverty, communities are socially dependent on governments, which also signifies the accountability role of municipalities, government and other organizations in addressing social vulnerability.

Prashar et al. [33] through a systemic review on urban flood resilience assessment agree that there is need for an integrated approach to achieve success in developing strong flood reliance. However, this study pointed out the absence of social vulnerability themes such as mental health, gender-based issues and perceptions within the flood resilience context. These themes are essential for recovery responses to residual risk. To maximize benefits to society, future responses to flooding events should consider contextual assessment of each case accounting for individual and integrated elements. Social dimensions are critical to the facilitation and rate of the recovery phase [33].

3.1.4 Risk perception

Risk perception includes assessing the symbolic and emotional relationships attached to place meanings [30]. This study reported on a study which included a survey of 707 residents in four coastal towns in France, South Africa and UK. However, the South African study shows a bias toward higher economic communities (retirees and international property owners). The coast of KZN for example has a very large population of lower economic dwellers as well. The outcomes of the study indicated varying flood risk perceptions that place significant meaning where natural features exist. However, residents should also be aware of the negative impacts such as flood risks in that location. The study has social implications especially where relocation as alluded to by Breen et al. [31] is considered as a strategic flood mitigation or response. A valuable contribution of the study by Quinn et al. [30] is the notions of "living with risks" and "protection against risks." The notion of "living with risks" involves socio-economic strategies such as insurance, timely communication of early warnings and other strategic responses such as moving furniture. This would be a huge challenge to marginalized poor socioeconomic communities such as in South African coastal towns, as they often do not have access to the resources.

Williams et al. [17] suggest the Capital Approach Framework (CAF). The approach with social capital as a component was applied to the Quarry Road West informal settlement, Durban, South Africa, within the context of informal settlements and flooding. This settlement went through periods of rapid expansion. Expansion was noted prior to and post significant flood events. Components of social capital include social networks, level of trust and confidence in stakeholders of decision-making and policy development. Five community members were among the stakeholder participants in the case study. This is significant in that the vulnerable communities were given a voice in the study. However, the results show that the vulnerable group felt they were not treated equally (0% score) in the governance of Quarry Road Settlement, with low scores for communication and participation as well. Equal and inclusive participation and co-design are essential for community buy-in and ownership of interventions and programs [35].

Neglect of social dimensions has contributed to flaws in the design of adaptation strategies as they disconnect from the lived experiences of individuals and communities (Adger et al., 2013, cited in [17]). Lived experiences contribute to understanding differences in vulnerability, as pointed out by Zhou et al. [32] in a climate change review in rural and urban areas. Residual risk of floods has the potential to exacerbate existing social vulnerabilities. The legacy of spatial racial segregation is still evident to date in areas such as Mitchell's Plain, Western Cape, South Africa where access to re-sources is limited [35]. Hence, localized interventions rather than "blanket policy" is valuable for adaptive strategies [32]. A reviewed input acknowledged the agency of women to adapt to floods but pointed out social vulnerabilities that challenged the agency more especially in rural areas *viz.* poverty and abuse [32].

The results of a study conducted in East London, South Africa, using the resilience theory as a framework to understand social vulnerability identified the elderly, people with special needs due to physical challenges, women and children in descending order as vulnerable. East London faces increasing flood risks to those already exposed to these events. Studying, analyzing and responding to social vulnerability of vulnerable groups are important for risk identification, preparedness, living through and recovering from hazardous events [28]. Understanding social vulnerabilities in flood risk areas to improve preparedness to events can be elevated to mandatory.

A case study conducted in the coastal Amanzimtoti community in Durban, South Africa, collected primary data (using questionnaires) from residents on lived experiences. This study highlighted social vulnerabilities post the 2019 floods [10]. Post the study, KZN has experienced recurring, damaging floods over the past 3 years. The study elaborates on both preparedness for flooding events and social impacts of previous flooding events. The study reports that: mitigation measures involved insurance policies and sandbag stacking that were eventually ineffective to protect assets in poor communities. Previous flooding events caused psychological trauma among some residents and other emotional challenges such as loneliness, fear and being isolated from usual religious gatherings. A fair share of the responsibility was expected from the municipality in terms of preparedness with residents, suggesting the provision of emergency tents, informed possible alternative relocation to churches and a subsidy availability for refurbishments would have done much for their well-being. Residents indicated that they were forced to temporarily relocate, sometimes in neighboring communities until flood water recession reached liveable conditions. However, residents seem to have positive place meaning and attachment to their town [30]. This deep attachment is evident in preparedness in adaptive interventions such as investment in sealant application on walls, flood detection systems, erecting flood barricades and insurance. On the contrary, some residents believed that risk management and accountability rested with the municipality and hence did not implement mitigation measures. Perhaps, an assumption on the part of this belief is that existing engineering infrastructure such as drainage was sufficient security. This highlights the importance of raising awareness among vulnerable communities of environmental, social and economic risks of complacency to flood preparedness. The lack of such awareness is to their detriment and is likely to escalate residual risk in communities with social vulnerabilities.

3.1.5 Communication and information dissemination

A common and chief concern that repeatedly appeared in this review was that of communication and information dissemination [10, 11, 17, 28]. Ineffective and incomprehensive communication and early warning systems exacerbate social vulnerability issues. For example, the anxiety and fear of vulnerable individuals, households and communities are likely to be confronted when events are suddenly upon them. Communication channels need to be timeously and proactively implemented, as television and radio news and social media warnings can be delayed in initiating urgent responses [11]. Further, electronic media works only when people are using them [28]. On the contrary Ebhuoma et al. [10] found that some residents preferred social media. However, they do concur that diverse communication mechanisms are more effective. In Bangladesh, coastal hazard inundation communication uses village networks and mosque authorities, using hand mikes and loudspeakers (Garai, 2017 cited in Sultana et al. [34]). Mosques are visible in many vulnerable communities in the South African coastal cities. Such community structures need to be explored as a co-strategy for communication during hazardous events. Respondents in the Amanzimtoti study conveyed the value of timeous communication and early warning indicating that it would allow time to protect or salvage key assets and make informed arrangements for relocation if necessary [10]. Williams et al. [17], raised the issue of communication and information post flood events in the Quarry Road West Settlement, citing lack of transparency in communication, the unavailability and inaccessibility of data on previous events, and the unavailability of reports in the languages of the residents as limitations, which needs to be considered in preparedness for future events.

The examination of social vulnerability in this review indicates that there is a dearth of published studies on the lived experiences of socially vulnerable coastal communities to flood risks, to inform increased preparedness and make recovery from residual risk meaningful and efficient. Further studies to address this gap have the potential to address residual risk more effectively, as the most vulnerable are yet to be fully understood within their contexts. Studies of lived experiences are also of value to communities across contexts where similarities may exist. Further, studies on lived experiences are needed and likely to provide a greater base for policy makers and practitioners to better prepare for flood hazards and recovery with an integrated approach. However, this review does categorically highlight the nuances around communication and early warning systems. To note though is the need for research on communication to include both early warning systems and real-time information to mitigate casualties and loss of assets [28]. Loss of life and its impact on future generations is one of the key social vulnerability issues of concern. This review favors policy, mitigation, preparedness and recovery founded on principles of integrated and hybrid approaches [25, 33], which includes social vulnerability rather than isolated focus on safe development engineering. The review also illuminates the number of approaches that can be applied by policy makers and practitioners to address social vulnerability and residual risk in an integrated way. More studies and publications on lived experiences of social vulnerability are likely to support (Sustainable Development Goal) SDG 11: *Make cities and human settlements inclusive, safe, resilient and sustainable within the context residual risk of flooding events.*

3.2 Economic implications

3.2.1 The significant impact of flooding on tourism

Floods in the Western Cape province of South Africa between 2011 and 2014 resulted in R682.8 million loss for municipalities, while floods in Mpumalanga province cost municipalities R535 million loss, and significant damage to infrastructure in Gauteng province due to floods between 2015 and 2018 [6]. South Africa's vulnerable coastal areas, which are economically dependent on tourism, have been significantly affected by ecological degradation and climate change [30, 36, 39]. The attrition of natural habitats impacts tourism by reducing the esthetic and recreational value of these regions, which in turn affects local livelihoods. The literature highlights that climate change has substantial economic implications for tourism, particularly in vulnerable regions like Southern Africa. The changing climate affects tourism activities due to its impact on the natural environment, which is a core attraction for many destinations [39, 40]. For instance, rising temperatures, changes in rainfall patterns, and increased frequency of EWEs can disrupt tourism activities and decrease the appeal of certain destinations, leading to reduced tourist arrivals and revenue losses. Coastal areas, which are popular for beach tourism, face threats from SLR and coastal erosion [41], increasing the economic vulnerability of these regions and affecting livelihoods reliant on tourism.

Estuaries are often prime locations for eco-tourism, with unique biodiversity and coastal landscapes that attract tourists. Van Niekerk et al. [42] reflect on the vulnerability of South African estuaries to climate change. The findings highlight the significant impacts of climate change on coastal ecosystems, particularly in the context of South Africa's diverse estuarine environments. More importantly, the vulnerability of estuarine areas is attributed to climate-induced stressors, including SLR, temperature changes, altered rainfall patterns, and more EWEs such as storms and floods. The vulnerability of estuaries has profound economic consequences for local communities, especially in areas reliant on these ecosystems for their livelihoods. Considering that South African estuaries provide critical services, including fisheries, tourism, and biodiversity, climate change can disrupt these services, threatening the economic stability of communities that depend on coastal resources. Tourism, particularly in areas where estuaries are a key attraction, is likely to face declines due to habitat loss, reduced biodiversity, and increased storm damage to coastal infrastructure. Van Niekerk et al. [42] proceed to argue that the impacts on tourism are closely linked to the degradation of estuarine habitats, associated with erosion of these sensitive ecosystems, particularly in regions like KZN. On a broader scale, the coastal nature of Durban means that tourism is a vital economic sector, heavily reliant on beach activities and ocean access. Flooding poses a direct threat to this sector by damaging infrastructure, such as beachfront facilities, and reducing tourist footfall during adverse weather conditions. The knock-on effect is a decline in revenue for local businesses that depend on tourism. Furthermore, flooding can result in environmental degradation, including beach erosion and pollution from overflowed sewage systems, which detracts from the city's attractiveness as a tourist destination [10, 17, 19, 32, 43].

The study by Fitchett et al. [44] focusing on the coastal towns of St. Francis Bay and Cape St. Francis can be aligned with the coastal areas of KZN Province, which also faces similar challenges related to climate change and coastal vulnerability. KZN's coast, popular for its thriving tourism sector, is highly susceptible to the economic impacts of climate change. As in the case of St. Francis Bay, rising sea levels, increased storm intensity, and changing weather patterns in KZN could threaten infrastructure, including beachfront properties, tourist attractions, and critical coastal services such as fishing and shipping. Tourism is a major economic driver in KZN, with destinations such as Durban, the iSimangaliso Wetland Park, and the Drakensberg Mountains attracting international visitors. The increased potential for coastal erosion and flooding could disrupt this industry, leading to significant revenue losses. In comparison, when aligning the findings from Dube, Nhamo, and Chikodzi's [45] study on SLR and coastal tourism in Cape Town, to the context of KZN Province, similar economic and environmental challenges emerge given the region's extensive coastline and reliance on coastal tourism. In KZN, the economic risks from rising sea levels parallel those identified in Cape Town, impacting key tourist destinations such as Durban's Golden Mile, Ballito, and Umhlanga. These areas draw substantial tourism income from beach activities, hotels, and beachfront properties. Rising sea levels and increased storm surges can result in significant infrastructure damage, increasing the costs of maintenance and repair. In KZN, tourism is a major employer, with coastal erosion or flooding resulting in job losses and reduced income for local communities, exacerbating poverty levels in already vulnerable populations. Furthermore, KZN's coastal zones include ecologically sensitive areas like estuaries and wetlands, which are critical for sustaining marine biodiversity and supporting eco-tourism initiatives. The loss of these habitats due to SLR can diminish the ecological value of the region, further affecting tourism potential.

Hence, the devastating implication of rising sea levels on economic growth opportunities, especially in coastal tourism, suggests the need for careful planning and investment in sustainable coastal infrastructure to protect both the economy and natural habitats, and to improve climate services to enhance disaster resilience in coastal communities by better aligning information with local needs, thereby reducing risks associated with extreme weather and flooding [40, 46, 47].

3.2.2 Infrastructure to support vulnerable communities and livelihoods

The literature interrogated by Dube et al. [45], Ebhuoma et al. [10], and Quin et al. [30] supports the arguments presented in the study conducted by Angelstam et al., which emphasizes the need for investments in maintaining and restoring ecological infrastructure to support the economic stability of vulnerable communities. This includes mitigating risks such as flooding and erosion that directly affect tourism and other coastal economic activities. The study "Mapping and evaluating the impact of flood hazards on tourism in South African national parks" by Dube et al. [45] offers a comprehensive analysis of how flood hazards affect the tourism sector in South Africa, particularly within national parks. The research underscores salient economic and social implications of flooding on vulnerable communities reliant on tourism as a primary source of income. In essence, flooding poses significant economic risks to the tourism sector, particularly in protected areas like national parks that attract international and local visitors. Flood events often result in infrastructure damage, such as washed-out roads, damaged visitor facilities, and increased maintenance costs [45]. The disruption to tourism activities leads to revenue losses, impacting both park authorities and the broader economy of surrounding communities, who depend heavily on tourism for their livelihoods [10, 19, 30, 35, 43]. Local businesses that provide ancillary services such as accommodation, food, and transport are impacted. This can lead to long-term economic instability in these regions, further marginalizing already vulnerable communities. Vulnerable communities living near estuaries may face displacement and loss of income as these impacts worsen. Furthermore, the shifting dynamics of marine currents and the effects on fish stocks can jeopardize local livelihoods, particularly those reliant on small-scale fishing [32, 34, 44, 48].

These arguments are strengthened by Ebhuoma et al. [10] and Williams et al. [17] within the context of flooding in Durban, South Africa. It is evident that economic impacts are significant, especially on low-income households in informal settlements. Williams et al. [17] outline the complexities of flood management in informal settlements, focusing on governance structures and their influence on the effectiveness of water management strategies. Poorly managed water governance can exacerbate the vulnerability of residents, leading to increased costs for infrastructure repair and loss of livelihoods. Economic activities in these areas are often informal and rely on fragile systems that are vulnerable to flood damage. Additionally, frequent flooding disrupts local economies, causing property damage, loss of assets, and displacement. For instance, in Quarry Road West (Durban, South Africa), poor governance and lack of infrastructure contribute to a cycle of poverty and increased economic vulnerability for residents. Residents often lack formal property rights, which limits their ability to advocate for necessary infrastructure improvements. This marginalization exacerbates their vulnerability to climate-induced flooding, and the absence of affordable insurance options exacerbates the financial strain on these communities. Consequently, economic losses manifest not only in property damage but also in disrupted livelihoods, as informal sectors such as street vending and small businesses face prolonged recovery periods after flood events. In addition, there is "business interruption and consequent reduced number of employed people, as well as transboundary economic losses and costs of emergency evacuation" [49]. This amplifies the need for financial safety nets and subsidies for vulnerable populations to mitigate the economic burden of climate-related disasters.

3.3 Application of appropriate risk management tools

The application of risk-based tools for the development of flood risk management strategies has largely featured the engagement of integrated models [13]. Such models consider trade-offs between management objectives and model uncertainties, to identify the critical parameters in each system. This has significant implications for refining and reconfiguring models as uncertain, unpredictable flood events occur. Multidisciplinary models consider ecology, economy, and society in measuring trade-offs against each other, to determine the consequences of alternatives under different scenarios [13].

Frazier et al. [21] applied a Residual Risk Assessment (RRA) model to assess risks to communities and health systems. This study highlighted the omission of health systems in many hazard assessments and recognized the risks and vulnerabilities of health systems. The RRA demonstrated the importance of such a model to help communities by determining residual risk and understanding how risks change over time. This can be used to prioritize risk reduction strategies and align community risk perception to be more reflective of reality and contribute to better risk communication. This can easily be applied specifically to flood events, as it allows for the RRA scores to "be compared across communities, the values can be compared between hazards and severities" [21].

Early warning systems (EWS) are important in risk communication to facilitate preparedness and response to climate risks [3]. EWS are best developed in collaboration with communities, as this will address the community needs. Often EWS are developed by government without consultation, resulting in it not achieving what it set out to do. Underpinning effective EWS is sound meteorological data. To obtain robust data, it is critical to maintain meteorological equipment in good order. This requires funding and adequately trained and skilled professionals to ensure that robust models are applied to generate data for the EWS. In the Southern African Development Community (SADC) region, resource challenges may create barriers for successful implementation of EWS [3]. Lunga et al. [50] conducted a study to determine an EWS based on a case study in Durban's Quarry Road informal settlement. This employed geographic information system (GIS) technology and spatial analysis of several criteria (including rainfall intensity, geology, land use, slope, and elevation) to identify flood-prone zones and contribute to EWS in the KZN region. South African Weather Services (SAWS) provide warnings associated with flash floods but do not provide levels of risk associated with vulnerable communities. This study concluded that there is a need for more intensive post-flood damage database, which would inform future predictive models. This requires investment to provide real-time spatial data to inform the EWS.

3.4 Consequences of future scenarios

Coastal floods resulting from storm events are projected to be more intense in combination with SLR, threatening storm protection infrastructure and coastal settlements. Coastal dune erosion reduces the buffering capacity, which offers protection to coastal infrastructure [41], requiring strategic planning to consider various possible future scenarios to guide decision making. The variables, introduced by human activities and natural phenomena, present coastal dynamics never considered before [14]. Quantitative global climate change projections produced in the United Nations Intergovernmental Panel on Climate Change (IPCC) Synthesis Report (Summary for Policymakers) are based on assumptions and a range of greenhouse gas (GHG) emission scenarios [4]. These include the following scenarios: very low and low GHG emissions, which are CO_2 emissions decreasing to net zero between 2050 and 2070; intermediate GHG emissions with CO_2 emissions approximately the same as current levels until 2050; and high and very high GHG emissions with approximately double current CO_2 emissions between 2050 and 2100. In the majority of gauged coastal locations, it is projected that the frequency of extreme sea level events will increase substantially. The risks to coastal communities will continue to rise beyond 2100 due to SLR associated with melting ice sheets and warming of the oceans. This will be exacerbated by continued increase in GHG emissions into the future, contributing to melting ice sheets in the Antarctic, with resultant SLR of 2–3 m in the next 2000 years if warming is limited to 1.5°C [4]. South Africa has experienced an annual SLR of approximately 1.71 mm since the 1990s, while the IPCC Synthesis report projects a global SLR of 0.18 m/annum for a low emissions scenario and 0.23 m/annum for a high emission scenario, for the year 2050 [20].

South Africa's coastline, of approximately 3751 km, includes the South Atlantic Ocean and the Indian Ocean, resulting in varied climatic conditions, environmental conditions, and coastal developments [20, 51]. Despite the variability, most of the coastline has something to offer, making it sought after. This variability results in regional differences, which would influence future projections, as there is no "one-size fits all." The implementation of adaptation plans linked to different scenario projections would also be subject to regional differences in the form of local government structures and resources. Johnston et al. [51] state that in a sustainability-focused scenario of 2°C rise in temperature, the eastern parts of SA are projected to have increased rainfall and reduced number of cyclones of higher intensity. SLR is set to increase, with a projected 90 cm increase with very high GHG emissions or a projected 50 cm increase in a very low GHG emission scenario. Despite efforts to reduce the impacts of climate change in coastal regions, the trajectory of negative consequences has been set in motion, but these impacts may be alleviated through mitigation and adaptation management strategies. Damages associated with SLR under a low-emission scenario may result in annual costs in the region of 306 million EUR or 815 billion EUR under high-emission scenario by 2050. In contrast, by 2100, the low-emission projected costs are 1.3 billion EUR and high-emission projection of 2.4 to 3.3 billion EUR [20]. This illustrates the alarming increase in costs associated with continuing with "business as usual."

High-emission scenarios are projected to have far-reaching consequences for the near to long-term future. Mitigation and adaptation measures to reduce such impacts of climate change on coastal communities will require significant planning and investment. Investment in such measures should link to the SDGs to achieve sustainable solutions for various scenarios, such as national and disaster risk reduction strategies. Climate change must be mainstreamed in national policies, strategies, and plans, along with general public awareness and capacity building in officials and decision-makers. South Africa currently has legislation covering climate change (Climate Change Act 22 of 2024) and disaster risk management (Disaster Management Act 57 of 2002), "aligned with international best practice" [52].

4. Implications for policy development

Considering the projected increase in climate-related impacts, there is a growing urgency for policies to protect tourism infrastructure and enhance the resilience of vulnerable communities. The adoption of sustainable tourism practices is encouraged [39, 40, 46] while exploring the implementation of collaborative governance involving multiple stakeholders, including local communities, industry players, and policymakers, is crucial for undertaking effective adaptation strategies. Angelstam et al. [36] advocate that enhancing collaborative governance and fostering partnerships between local communities, policymakers, and private sectors is key strategy proposed to drive sustainable development and adaptation measures. The authors recommend active adaptive management, which involves iterative learning and evidence-based policy adjustments to respond to ongoing environmental changes effectively.

The literature by Claassens et al. [46], Cooper and Green, [40] and de Wit et al. [48] underscore the necessity for comprehensive policy frameworks that address the cumulative effects of coastal development. There is a critical demand for baseline data to inform policy decisions, guiding future infrastructure projects in a manner that balances development and ecological sustainability, through integrated coastal zone management (ICZM) approaches and adaptive coastal management policies. Van Niekerk et al. [42] affirm the importance of integrated coastal zone management that considers both the socioeconomic and environmental aspects of estuary management, therefore, the call for policies that focus on climate resilience, such as coastal protection measures, habitat restoration, and sustainable resource management practices. These include integrating coastal infrastructure planning with climate adaptation planning strategies, promoting sustainable tourism practices, and considering nature-based solutions such as dune restoration to mitigate erosion, and policies that strengthen the capacity of local communities to adapt to changing environmental conditions [19, 40, 42, 46, 48]. Policies need to focus on promoting sustainable development practices that incorporate climate resilience, particularly in informal settlements where the urban poor often live. The outcomes of the study by Tauhid and Zawani [19] stress the need for policies to integrate green infrastructure into urban planning. For example, infrastructures, such as green roofs, urban wetlands, and permeable pavements, help absorb and manage stormwater, reducing flood risks.

Policymakers are encouraged to promote community participation, ensuring that the solutions are tailored to local needs and are sustainable over time. In effect, the policy development process must prioritize community participation, integrating local knowledge and fostering multi-sector partnerships. This implies that local governance structures need to be more inclusive and capable of addressing the needs of vulnerable communities. Williams et al. [17] advocate for the involvement of residents in flood risk management, as they possess crucial knowledge about local vulnerabilities that can inform better decision-making [32]. On the contrary, Fitchett et al. [44] assert that local perceptions of climate risks are often disconnected from scientific predictions. For instance, many accommodation proprietors are aware of the threats posed by climate change, but there is limited proactive adaptation, with responsibility frequently placed on the government, despite its capacity limitations. This gap between perceived and actual risk highlights the need for more robust policy frameworks that integrate scientific data with local perceptions and provide tangible climate-resilient adaptation strategies. According to Goble et al. [53] and van Niekerk et al. [42], the coastal management, in South Africa, has evolved drastically, moving from sector-based management to an integrated and holistic approach. The government has implemented frameworks like the National Coastal Management Program (NCMP) to better address coastal risks, including those linked to climate change and flooding. This has led to improved policy development aimed at reducing vulnerabilities in coastal communities, especially in the face of rising sea levels.

Policy development initiatives highlight the need for interdisciplinary approaches, involving local governments, communities, and environmental experts. Strengthening institutional capacities, improving flood resilience, and incorporating environmental and social justice into tourism planning are critical for long-term sustainability. The systematic review by Sultana et al. [34] underscores the integration of local adaptation practices into formal policy frameworks, promoting risk-informed nature-based solutions and community-driven initiatives [54]. The underlying principle is that policies need to be adaptive and flexible, addressing both short-term impacts and long-term risks associated with climate change. Additionally, policies must recognize the social values and cultural practices of informal communities, which can enhance adaptive capacity and resilience [35].

Studies by Dube et al. [45], Ebhuoma et al. [10], and Quinn et al. [30] advocate for a more integrated approach to policy development, focusing on proactive flood risk management strategies. They suggest that tourism policies should incorporate climate adaptation measures to protect infrastructure and ensure the safety of tourists. This includes early warning systems, "investment in climate-resilient infrastructure, transparent management of public resources" [54], and comprehensive disaster response plans. Ebhuoma et al. [10] explore the various structural and non-structural measures applied by eThekwini Municipality to enhance flood resilience. However, the literature raises barriers to communication as an enduring gap to effective early warning systems. For instance, during the 2019 floods, the failure to use preferred communication channels like social media reduced the effectiveness of early warning systems. To improve flood resilience, the study recommends enhancing community engagement, particularly with marginalized groups, to tailor interventions effectively. The adoption of inclusive communication strategies using a mix of traditional media (radio, TV) and digital platforms (social media) could increase the reach and effectiveness of early warnings. The authors also support the notion of integrated policy frameworks that align disaster risk reduction with broader urban development goals, contributing to the achievement of Sustainable Development Goal 11, which aims for inclusive, safe, resilient, and sustainable cities. In essence, addressing the economic and social impacts of flooding in Durban requires a multi-faceted approach that includes improving infrastructure, enhancing community preparedness, and developing targeted policies that support vulnerable populations. In addition, transparent risk communication is critical in preparing adequately for management of residual risks [23].

Quinn et al. [30] investigate on how community attachments to local waterbodies, such as rivers and lakes, influence flood risk perception and management preferences. It reviews the role of symbolic and emotional connections in shaping risk decisions. People's attachment to landscapes, including economic and cultural ties, can hinder preparedness and risk management efforts, especially in high-risk areas. Policymaking should integrate local knowledge and perceptions to enhance community engagement and the acceptance of flood management measures. Regarding policy implications to address rising sea level risk factors, Dube et al. [55] report that policies should focus on sustainable adaptation measures, including the construction of seawalls, beach nourishment, and the restoration of natural barriers such as dunes and wetlands. Equally important is taking strategic actions in enhancing community awareness and involvement in climate adaptation planning to ensure the resilience of the tourism sector. Engaging local communities in adaptation planning and providing education on the impacts of climate change can help align public efforts with sustainable tourism practices. There is also a need for better data collection and monitoring systems to effectively predict and respond to sea level changes. Some of the barriers to effective climate governance were explored in a study by Adom et al. [7]. These included lack of funding, poor investment in climate change, and poorly trained, inadequately qualified and skilled professionals to implement policies and programs to address climate governance issues. This requires adequate resources and funding to be addressed as part of adaptation planning and residual risk management.

5. Residual risk management and the SDGs

Risks associated with flooding may seriously compromise progress toward achieving the sustainability agenda. Flood risk associated with vulnerable communities exacerbates the current dire situation in many countries in the Global South, where poverty, food insecurity, and compromised wellbeing are part of daily life. Hence, it is critical that potential risks that persist beyond the implementation of protection measures are managed to reduce further suffering and loss. Climate change adaptation is critical in achieving the SDGs, but reducing the impacts of flooding, among other climate risks, requires planning [10]. This may require finding opportunities in potential negative impacts and being proactive in implementing activities, policies, and programs to reduce negative impacts of floods. **Table 1** below explores the opportunities that are linked to the negative consequences associated with flood events.

SDG No.	SDG description	Consequences of flood events and associated opportunities to reduce residual risk and contribute to the SDGs	
		Negative consequences	*Opportunities to contribute toward achieving the SDGs*
1	No poverty	Loss of livelihoods, compounded by forced temporary or permanent displacement.	Development and implementation of policies that address mitigation measures to protect vulnerable communities.
3	Good health and wellbeing	Diseases due to sewage infrastructure failures.	Strengthen health systems to manage residual risks early; training of personnel; ensuring adequate resources are available, which may provide additional employment opportunities.
6	Clean water and sanitation	Extensive damage to sewage and water infrastructure compromises water quality.	Equip communities with more robust infrastructure or temporary services in the event of infrastructure being compromised; create awareness among communities relating to these issues, especially impact of water quality on health.
8	Decent work and economic growth	Impacts on livelihoods and economy.	Work with communities to establish the possible entrepreneurial opportunities, such as nature-based solutions to prevent coastal erosion.
9	Industry innovation and infrastructure	Damage to infrastructure.	Opportunity for the development of adaptation measures to reduce residual risk; development of ecological infrastructure [36].
10	Reduced inequalities	Loss of livelihoods; lack of resources to recover timeously.	Partnerships to bridge the divide facilitating coping mechanisms, linked to SDG17. Government budget allocation to address recovery from loss of livelihoods

SDG No.	SDG description	Consequences of flood events and associated opportunities to reduce residual risk and contribute to the SDGs	
		Negative consequences	*Opportunities to contribute toward achieving the SDGs*
11	Sustainable cities and communities	Migration to urban areas and settlement in unsuitable landscapes [8].	Restoration of functional ecological infrastructure [36]. Provision of more robust building materials or robust dwelling design to replace unsafe informal dwellings.
13	Climate action	Social, environmental, and economic sectors' resilience to climate adaptation compromised.	Use lessons learned to revise adaptation planning and improve mitigation measures.
14	Life below water	Coastal ecosystems disrupted.	Use of nature-based solutions to ensure ecosystem resilience to climate change. Dune stabilization and regulations to control coastal developments.
15	Life on land	Terrestrial ecosystems disrupted and resilience to further disruptions compromised.	Investing in ecological infrastructure to maintain resilient ecosystems and provide ecosystem services.
16	Peace and justice strong institutions	Contravention of the rule of law results in people settling in vulnerable environments [8].	Creating a wider awareness of climate change and the activities that contribute to exacerbated flood risk associated with climate change.
17	Partnerships for the Goals	Inadequate budget allocation and resources for flood disasters at different tiers of government, and across government and private sectors.	More partnerships required to ensure faster recovery period and pooling of resources.

Table 1.
Residual risk may be managed to influence the extent of negative consequences of flood events and to enhance the potential opportunities toward achieving the SDGs.

6. Lessons learned, conclusion, and recommendations

The resounding message from Grab and Nash [9] in their study on KZN indicated that the scale of the floods experienced in KZN, South Africa, in 2022, is likely to recur and possibly be of a larger magnitude. Although the aid provided by government and non-government organizations was significant, it would be preferable to reduce the suffering in the aftermath of such EWEs. Hence, proactive planning is critical to manage and reduce residual risks. In summary, the salient lessons learned and recommendations on flood risk management policy development drawn from the literature reviewed include the necessity of promoting transdisciplinary knowledge production and social learning among diverse stakeholders to bridge existing gaps in understanding and action. Strengthening resilience depends heavily on collaboration across sectors, from disaster risk management to sustainable tourism practices. The authors Angelstam et al. [36] and Goble et al. [53] advocate for scaling up research and development projects that integrate ecological and social insights, facilitating investments in green infrastructure. Practical recommendations involve prioritizing conservation and landscape restoration initiatives that align with broader sustainability goals. By fostering collaboration across different scales and sectors, authorities can better manage residual risks and enhance the resilience of vulnerable communities against climate-related impacts.

Claassens et al. [46] heighten the focus on regulatory frameworks to oversee coastal developments and prevent *ad hoc* infrastructure expansion, which can lead to significant environmental degradation. Applying stricter environmental impact assessments (EIAs) and encouraging eco-friendly engineering solutions, such as living shorelines, could enhance coastal resilience. Another crucial action is the initiation and maintenance of a detailed inventory of existing infrastructure to monitor changes and guide future developments. This approach could help identify areas most at risk and prioritize interventions that balance economic growth with environmental conservation, while considering potential residual risks post-flooding.

In further enhancing coastal monitoring and urban resilience planning [40, 48, 55], justifying the importance of investing in sustainable infrastructure, and fostering community involvement in resilience planning through a multi-dimensional approach [23], integrating economic, social, and natural capital considerations, is critical. The understanding is that prioritizing ecological conservation and promoting multi-stakeholder collaboration can protect coastal economies and ecosystems against future climate impacts. Collaborative efforts between government bodies, the tourism industry, and local communities [23] to effectively manage and mitigate the impacts of SLR warrant much attention, stressing the need for long-term monitoring of coastal changes and the implementation of early warning systems to minimize economic disruptions. Equally important is for policy makers to reinforce local governance mechanisms, incorporating climate risk assessments into adaptive urban planning, and fostering stakeholder collaboration in driving climate resilience. This is coherently endorsed by Dube [39] in the description of adaptive management in tourism planning. Proactive measures, such as diversifying tourism products and services, can help reduce dependence on vulnerable coastal areas. Additionally, the integration of climate change education into tourism planning can raise awareness and encourage the adoption of more sustainable practices. Dube recommends that future research should focus on identifying specific adaptation strategies tailored to the unique needs of different tourism destinations, particularly in regions highly susceptible to climate impacts. In effect, there needs to be a deliberate shift in how tourism is managed, providing for greater consideration of climate resilience in tourism policies to safeguard economic interests and support sustainable development. The outcomes of the study undertaken by Dube et al. [55] propose collaborative efforts between government bodies, the tourism industry, and local communities to effectively manage and mitigate the impacts of SLR.

In the same vein, Dube et al. [45] postulate enhancing collaborative efforts between disaster management authorities, park management, and local communities to build resilience, through integration of stakeholder engagement in decision-making processes to develop sustainable tourism strategies that mitigate flood risks and reduce possible residual risks. Addressing the economic and social impacts of flooding in South Africa requires a multi-faceted approach that includes improving infrastructure, enhancing community preparedness, and developing targeted policies that support vulnerable populations [10, 44]. Quinn et al. [30] clarify the importance of integrating local knowledge into flood risk management to foster effective, community-driven, and culturally responsive policies. As highlighted by

Ebhouma et al. [10], mental health of communities following flood events requires attention, to help them cope with the devastation, loss, and displacement. This embraces a more holistic approach to residual risk management. The authors espouse for a paradigm shift in flood management strategies, from relying solely on traditional infrastructure to incorporating nature-based solutions [6]. This approach not only mitigates flooding but also enhances urban resilience, biodiversity, and social equity, making it a multifaceted strategy for sustainable urban development. The lessons learned underscore the importance of collaborative governance, adaptive policies, and investments in green infrastructure to address the complexities of urban flooding, especially in vulnerable communities. Strengthening institutional capacities, improving flood resilience, incorporating scientific projections, local knowledge, and practical adaptation strategies to safeguard and sustain human well-being, tourism, and coastal livelihoods are vital in preparing for future disasters as we anticipate an increase in EWEs.

Acknowledgements

The authors acknowledge the contribution of Research Interns, Nondumiso Mokwena and Maphili Khumalo in the initial literature searches, whereafter the authors refined and screened the results and analyzed the literature.

Author details

Indrani Hazel Govender[1]*, Maliga Reddy[2] and Rajendran Perumal Pillay[3]

1 Department of Horticulture, Durban University of Technology, Durban, South Africa

2 Department of Public Management and Economics, Durban University of Technology, Durban, South Africa

3 3H Enviro Strategies Consulting, Durban, South Africa

*Address all correspondence to: indranig@dut.ac.za; hazeyg@manaar.co.za

References

[1] Kron W. Flood risk hazard values vulnerability. Water International. 2005;**30**(1):58-68

[2] CDRF. Climate & Disaster Resilience Fund. South Africa: Climate & Disaster Resilience Fund; 2024. Available from: https://climateresiliencefund.org.za/

[3] Agbehadji IE, Schütte S, Masinde M, Botai J, Mabhaudhi T. Climate risks resilience development: A bibliometric analysis of climate-related early warning Systems in Southern Africa. Climate. 2023;**12**(1):3

[4] IPCC. Summary for policymakers. In: Core Writing Team, Lee H, Romero J, editors. Climate Change 2023: Synthesis Report. Contribution of Working Groups I, II and III to the Sixth Assessment Report of the Intergovernmental Panel on Climate Change. Geneva, Switzerland: IPCC; 2023

[5] Wang H-W, Castro DS, Chen GW. Managing residual flood risk: Lessons learned from experiences in Taiwan. Progress in Disaster Science. Jun 2024:100337

[6] CSIR. Flooding: Current state and implications of climate change on flooding in South Africa. In: GreenBook National Overview. Pretoria, South Africa: Council for Scientific and Industrial Research; 2019. Available from: https://pta-gis-2-web1.csir.co.za/portal/apps/GBCascade/index.html?appid=33d9a846cf104e1ea86ba1fa3d197cbd

[7] Adom RK, Simatele MD, Das DK, Mukalazi KA, Sonwabo M, Mudau L, et al. Enhancing climate change adaptation governance through transforming institutions in Kwa-Zulu Natal Province, South Africa. International Journal of Climate Change Strategies and Management. 2024;**16**(4):413-438

[8] Dube K, Nhamo G, Chikodzi D. Flooding trends and their impacts on coastal communities of Western Cape Province, South Africa. GeoJournal. 2022;**87**(Suppl. 4):453-468

[9] Grab SW, Nash DJ. A new flood chronology for KwaZulu-Natal (1836-2022): The April 2022 Durban floods in historical context. South African Geographical Journal. 2024;**106**(4):476-497

[10] Ebhuoma EE, Nene NJ, Leonard L. Analysis of urban households' preparedness and municipal interventions to build flood resilience in Durban, South Africa: Implications for SDG 11. Environmental and Sustainability Indicators. 2024;**23**:100454

[11] Mudefi E. Disaster management 'deeds' in the context of April 2022 KwaZulu-Natal floods: A scoping review. International Journal of Disaster Risk Reduction. 2023;**98**:104122

[12] Aswani S, Howard J, Gasalla MA, Jennings S, Malherbe W, Martins I, et al. An integrated framework for assessing coastal community vulnerability across cultures, oceans and scales. Climate and Development. 2019;**11**(4):365-382

[13] Oddo PC, Lee BS, Garner GG, Srikrishnan V, Reed PM, Forest CE, et al. Deep uncertainties in sea-level rise and storm surge projections: Implications for coastal flood risk management. Risk Analysis. 2020;**40**(1):153-168

[14] Cartwright A. Coastal Vulnerability in the Context of Climate Change:

A South African Perspective. Cape Town: Centre for Criminology; 2011

[15] Cea L, Costabile P. Flood risk in urban areas: Modelling, management and adaptation to climate change. A review. Hydrology. 2022;**9**(3):50

[16] Welisiejko S, Cáceres B. Informal Settlements: No Longer Invisible. The Role of Impact in Scaling Capital Mobilisation to Fund Slum-Upgrading Programmes Globally. London, UK: The Global Steering Group for Impact Investment; 2022

[17] Williams DS, Máñez Costa M, Celliers L, Sutherland C. Informal settlements and flooding: Identifying strengths and weaknesses in local governance for water management. Water. 2018;**10**(7):871

[18] Lunga W, Nkosi M, Tirivanhu P, Madzivhandila T, Ramaphakela T, Majikijela Y, et al. Strengthening disaster risk resilience through enhanced coordination mechanisms at local level in flood hazard prone areas: Case study of KwaZulu-Natal province. In: Pretoria, South Africa: HSRC DCES Report. HSRC; 2023

[19] Tauhid F, Zawani H. Mitigating climate change related floods in urban poor areas: Green infrastructure approach. Journal of Regional and City Planning. 2018;**29**:98

[20] CMCC. G20 Climate Risk Atlas. Impacts, Policy, Economics. South Africa: CMCC; 2024. Available from: https://www.g20climaterisks.org/south-africa/

[21] Frazier TG, Wood EX, Peterson AG. Residual risk in public health and disaster management. Applied Geography. 2020;**125**:102365

[22] RSA. Climate Change Act 22 of 2024. Government Gazette No 50966. Pretoria, South Africa: Parliament of the Republic of South Africa; 2024

[23] Hartmann S, Pedoth L, Dalla Torre C, Schneiderbauer S. Beyond the expected—Residual risk and cases of overload in the context of managing alpine natural hazards. International Journal of Disaster Risk Science. 2021;**12**(2):205-219

[24] UNDRR. Residual Risk: Switzerland: United Nations Office for Disaster Risk Reduction. 2024. Available from: https://www.undrr.org/terminology/residual-risk

[25] Busayo ET, Kalumba AM, Afuye GA, Olusola AO, Ololade OO, Orimoloye IR. Rediscovering South Africa: Flood disaster risk management through ecosystem-based adaptation. Environmental and Sustainability Indicators. 2022;**14**:100175

[26] Bwambale B, Nyeko M, Muhumuza M, Kervyn M. Questioning knowledge foundation: What is the best way to integrate knowledge to achieve substantial disaster risk reduction? International Journal of Disaster Risk Reduction. 2020;**51**:101850

[27] Ajtai I, Ştefănie H, Maloş C, Botezan C, Radovici A, Bizău-Cârstea M, et al. Mapping social vulnerability to floods. A comprehensive framework using a vulnerability index approach and PCA analysis. Ecological Indicators. 2023;**154**:110838

[28] Busayo ET, Kalumba AM. Recommendations for linking climate change adaptation and disaster risk reduction in urban coastal zones: Lessons from East London, South Africa. Ocean & Coastal Management. 2021;**203**:105454

[29] Lockwood C, Munn Z, Porritt K. Qualitative research synthesis: Methodological guidance for systematic reviewers utilizing meta-aggregation. JBI Evidence Implementation. 2015;**13**(3):179-187

[30] Quinn T, Bousquet F, Guerbois C, Heider L, Brown K. How local water and waterbody meanings shape flood risk perception and risk management preferences. Sustainability Science. 2019;**14**:565-578

[31] Breen MJ, Kebede AS, König CS. The safe development paradox in flood risk management: A critical review. Sustainability. 2022;**14**(24):16955

[32] Zhou L, Kori DS, Sibanda M, Nhundu K. An analysis of the differences in vulnerability to climate change: A review of rural and urban areas in South Africa. Climate. 2022;**10**(8):118

[33] Prashar N, Lakra HS, Shaw R, Kaur H. Urban flood resilience: A comprehensive review of assessment methods, tools, and techniques to manage disaster. Progress in Disaster Science. Oct 2023:100299

[34] Sultana T, Islam MR, Mustafa FB, Sim JOL. A systematic review of coastal community adaptation practices in response to climate change-induced tidal inundation. Journal of Coastal Conservation. 2022;**26**(4):32

[35] Mclachlan J, Tanyanyiwa CT, Schneuwly R, Carden K, Armitage NP, Abrams A, et al. Pathways to water resilient south African cities–from mono-functional to multi-functional stormwater infrastructure. Scientific African. 2023;**20**:e01674

[36] Angelstam P, Barnes G, Elbakidze M, Marais C, Marsh A, Polonsky S, et al. Collaborative learning to unlock investments for functional ecological infrastructure: Bridging barriers in social-ecological systems in South Africa. Ecosystem Services. 2017;**27**:291-304

[37] Kamara JK, Akombi BJ, Agho K, Renzaho AM. Resilience to climate-induced disasters and its overall relationship to well-being in southern Africa: A mixed-methods systematic review. International Journal of Environmental Research and Public Health. 2018;**15**(11):2375

[38] Muyambo F, Belle J, Nyam YS, Orimoloye IR. Climate-change-induced weather events and implications for urban water resource management in the free state province of South Africa. Environmental Management. 2023;**71**(1):40-54

[39] Dube K. Evolving narratives in tourism and climate change research: Trends, gaps, and future directions. Atmosphere. 2024;**15**(4):455

[40] Cooper JAG, Green AN. Southern African sandy coasts in the context of near-future sea-level rise. Transactions of the Royal Society of South Africa. 2023;**78**(3):149-166

[41] Guastella LA, Rossouw M. Coastal vulnerability: What will be the impact of increasing frequency and intensity of coastal storms along the south African coast. Reef Journal. 2012;**2**:129-139

[42] Van Niekerk L, Lamberth SJ, James NC, Taljaard S, Adams JB, Theron AK, et al. The vulnerability of south African estuaries to climate change: A review and synthesis. Diversity. 2022;**14**(9):697

[43] Rodina L, Harris L, Ziervogel G, Wilson J. Resilience counter-currents: Water infrastructures, informality, and inequities in Cape Town, South Africa. World Development. 2024;**180**:106619

[44] Fitchett JM, Grant B, Hoogendoorn G. Climate change threats to two low-lying south African coastal towns: Risks and perceptions. South African Journal of Science. 2016;**112**(5-6):1-9

[45] Dube K, Nhamo G, Chikodzi D, Chapungu L. Mapping and evaluating the impact of flood hazards on tourism in south African national parks. Journal of Outdoor Recreation and Tourism. 2023;**43**:100661

[46] Claassens L, de Villiers N, Waltham N. How developed is the south African coast? Baseline extent of South Africa's coastal and estuarine infrastructure. Ocean & Coastal Management. 2022;**222**:106112

[47] Lumbroso D, Vincent K, Murambadoro M, Steynor A, Tsarouchi G, Nezi M. Current uses and potential future needs for climate services in South Africa. Climate Services. 2024;**36**:100516

[48] de Wit M, Rawlins J, Petrie B. Economic risk assessment of climate change at the city level. The case of Cape Town, South Africa. International Journal of Urban Sustainable Development. 2023;**15**(1):118-140

[49] Tanoue M, Taguchi R, Alifu H, Hirabayashi Y. Residual flood damage under intensive adaptation. Nature Climate Change. 2021;**11**(10):823-826

[50] Lunga W, Nkosi M, Chirima G, Madzivhandila T, Ratshiedana P, Ramaphakela T, et al. Developing an Early Warning System to Strengthen Disaster Risk Resilience: Experiences from Flood-Prone Areas of KwaZulu-Natal Province. Research Square Preprint; 2023

[51] Johnston P, Egbebiyi TS, Zvobgo L, Omar SA, Cartwright A, Hewitson B. Climate Change Impacts in South Africa: What Climate Change Means for a Country and its People. South Africa: University of Cape Town; 2024

[52] STATSSA. Sustainable Development Goals Country Report 2023. South Africa. Pretoria, South Africa: Statistics South Africa; 2023

[53] Goble BJ, Lewis M, Hill TR, Phillips MR. Coastal management in South Africa: Historical perspectives and setting the stage of a new era. Ocean & Coastal Management. 2014;**91**:32-40

[54] Muyambo F, Belle J, Nyam Y, Orimoloye I. Building resilience to multiple climate change-related risks in QwaQwa using the community capitals approach. Journal of Water and Climate Change. 2024;**15**(8):3431-3449

[55] Dube K, Nhamo G, Chikodzi D. Rising sea level and its implications on coastal tourism development in Cape Town, South Africa. Journal of Outdoor Recreation and Tourism. 2021;**33**:100346